LIVING BY CHEMISTRY
Preliminary Edition

PROGRAM SAMPLER

Angelica M. Stacy
College of Chemistry
University of California at Berkeley

with

Jan Coonrod

Jennifer Claesgens

 **Key Curriculum Press**

Project Editor: Ladie Malek

Editorial Assistant: Elizabeth Ball

Consultant: Brian Erwin

Production Director: McKinley Williams

Production Editor: Christine Osborne

Copyeditor: Jacqueline Gamble, Angela Chen

Production Coordinator: Thomas Brierly

Text Designer: Charice Silverman

Compositor: Charice Silverman, Thomas Brierly

Art Editor: Jason Luz

Cover Designer: Jensen Barnes

Back Cover Photo Credits: *Left:* © M. J. Wickham;
Center: RF; *Right:* © M. J. Wickham

Prepress and Printer: Data Reproductions

Sales Director: Kelvin Taylor

Executive Editor: Casey FitzSimons

Publisher: Steven Rasmussen

Acknowledgments

Living By Chemistry has been developed at the University of California at Berkeley. This material is based upon work supported by the National Science Foundation grant No. ESI-9730634. Any opinions, findings, and conclusions or recommendations expressed in this material are those of the author and do not necessarily reflect the views of the National Science Foundation or the Regents of the University of California.

Published by Key Curriculum Press

Key Curriculum Press
1150 65th Street
Emeryville, CA 94608
510-595-7000
editorial@keypress.com
www.keypress.com

Printed in the United States of America

10 9 8 7 6 5 4 3 2 1   10 09 08 07 06 05   Part No.: 87796

Contents

LIVING BY CHEMISTRY PROGRAM SAMPLER
©2006 Key Curriculum Press

INTRODUCTION

Sampler Overview

This sampler is provided to familiarize you with the approach, content, and features of *Living By Chemistry*, a high school–level general chemistry curriculum now available in a preliminary edition. Both teacher and student material from selected sample lessons are included. We invite you to try these lessons with students in your school or district, and to share your thoughts with us. Contact us at editorial@keypress.com with comments or questions.

To request complete examination copies of the units after reviewing this sampler, please call (800) 995-6284.

Overview of *Living By Chemistry* Program

BREAKTHROUGH CURRICULUM

Living By Chemistry is a high school chemistry curriculum created by Professor Angelica Stacy and her team of developers at the University of California at Berkeley under a grant from the National Science Foundation. The goal of the *Living By Chemistry* program is to provide an inquiry-based, yearlong high school chemistry course that surpasses both national and state chemistry standards to improve science understanding for all students. The program provides groundbreaking materials for educators who seek a chemistry program that is content-rich, yet accessible to students.

Students learn best when they are actively engaged in investigating questions in an interesting context. To promote subject matter coherence for students, *Living By Chemistry*, Preliminary Edition, is organized into five units, each unit covering one of the big ideas of chemistry. Each unit is centered on a contextual theme. The context holds the interest of the students and provides a real-world foundation for the chemistry concepts.

Unit 1—*Alchemy: Atoms, Elements, and the Periodic Table*

What is the make-up of the world around us? Students explore the structure of the atom through a historical lens, learning about matter, atoms and elements, the periodic table, electron configuration, and bonding. They follow in the footsteps of scientists to re-create the modern periodic table and evaluate atomic models in light of experimental evidence.

Unit 2—*Smells: Molecular Structure and Properties*

How does the nose know? Students investigate the chemistry of smell, learning about molecular formulas, Lewis dot structures, bonding, and the shape and structure of molecules.

Unit 3—*Weather: Gas Laws and Phase Changes*

What's the forecast? Students ask questions related to weather, leading to an understanding of specific heat capacity and phase changes, and of the relationships among pressure, temperature, and volume that lead to the gas laws.

Unit 4—*Toxins: Chemical Reactions and Stoichiometry*

How much is too much? Students explore the concept of toxicity, learning about the qualitative and quantitative aspects of chemical reactions, including stoichiometry, balancing equations, acid-base chemistry, and precipitation reactions.

Unit 5—*Fire: Energy and Thermochemistry*

What is the nature of fire? Students investigate the intimate connection between change and energy, exploring thermodynamics and activation energy. This unit covers heat transfer, combustion, bond energies, oxidation-reduction, and enthalpy changes.

LIVING BY CHEMISTRY PROGRAM SAMPLER
©2006 Key Curriculum Press

SUCCESS FOR MORE STUDENTS

Chemistry is a gateway course to numerous attractive careers, particularly careers in the health sciences and engineering. Therefore, we want students to succeed in chemistry and have the option to consider these careers. Judging by past outcomes, if chemistry continues to be taught conventionally, using compendium-style texts, many students will be frustrated and will fail to meet the course expectations. Even students who receive high marks under current practices do not necessarily gain a working knowledge of chemistry or appreciate the relevance of its key ideas. Research has shown that many more students can learn the subject matter and learn the subject matter in depth, if we expand the pedagogical approach used to teach the subject, using a diversity of carefully constructed lessons that are both engaging and instructive. The result is a chemistry course both content-rich and highly accessible.

Living By Chemistry has been developed and tested in diverse chemistry classrooms. It is currently in use in large and small districts in several states, as well as in private and charter schools. Our experience so far indicates that all students can be engaged and gain a good understanding of the material through this guided inquiry approach. Pre- and post-tests used in field-test classrooms show that this innovative approach benefits all students.[1]

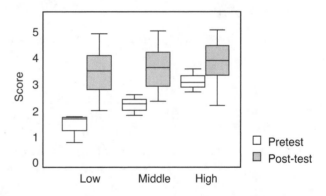

Teachers using this program can rest assured that students are making progress toward national standards and gaining a deep understanding of chemistry, while also experiencing the excitement of a hands-on, inquiry-based approach to the science of chemistry.

ACTIVE PARTICIPATION

Each unit consists of real-world investigations that encourage active student participation. In the second unit, for example, students investigate why different substances have different smells. All ideas and explanations are considered as students continually refine their ideas through activities and discussions.

---

[1] From "Living by Chemistry: Evidence of Student Learning Gains" by Jennifer Claesgens, et al. © UC Regents, Living By Chemistry 2003. Copies are available by request from Key Curriculum Press.

Each lesson opens with a ChemCatalyst, a question (or questions) designed to elicit students' ideas about what they are about to learn. After a teacher-guided discussion, the class moves into the daily Activity. These Activities are carefully chosen to illustrate the most difficult ideas for students, and take different forms depending on the best vehicle for student understanding of the subject matter: lab experiments, model building, card sorts, card games, group problem solving, or worksheets with review problems. Students often work in groups of two or four. The variety of activities ensures that different learning modalities are addressed and that learning is reinforced in several ways for each student.

Living By Chemistry encourages active engagement with the subject matter. Studies have shown that this engagement augments student learning, and increases the likelihood that students will enjoy the course and choose to continue their education in science.

Living By Chemistry teaches rigorous chemistry content. Students perform laboratory experiments to form new knowledge, not merely to confirm facts learned in class. Students also sort and organize information, build molecular models, and evaluate the merits and shortcomings of physical models for explaining phenomena. In short, the way students learn chemistry using *Living By Chemistry* reflects the way chemists carry out their work.

GUIDED INQUIRY

The role of the teacher is key to student success in any classroom. Teachers ask the questions that further students' understanding, help students synthesize what was learned each day, and introduce content and definitions when they are most appropriate and helpful to student learning.

Guiding students effectively is made easy by the lesson structure of *Living By Chemistry*. All the pieces of the lesson are laid out for optimal lesson flow, and ample background information is provided in the Teacher Guides so that teachers feel comfortable guiding the flow of learning.

The *Living By Chemistry* Teacher Guides offer unique support to teachers through daily lesson plans based on a constructivist view of teaching and learning. These include a lesson opener, an activity, relevant content including definitions and diagrams, and questions to ask along the way, with sample student answers included. The Teacher Guides also contain all necessary reproducible pages: worksheets, handouts, transparency masters, and homework assignments.

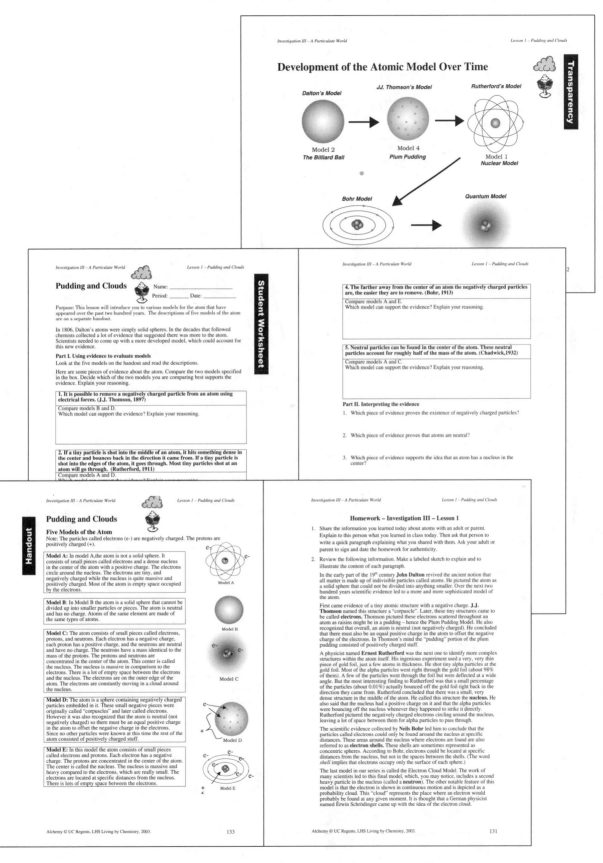

The lessons adhere to a predictable format, as shown in the following pages.

Lesson Structure

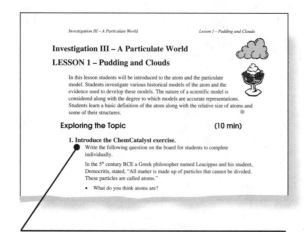

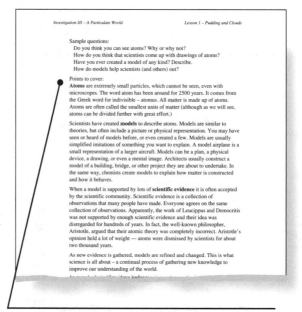

Each lesson opens with a ChemCatalyst. This lesson opener engages students right away and encourages a high level of participation from all students.

Relevant background information is provided for teachers to share with students.

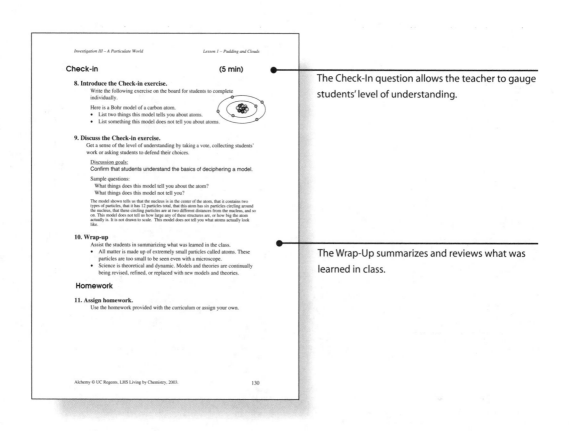

The Check-In question allows the teacher to gauge students' level of understanding.

The Wrap-Up summarizes and reviews what was learned in class.

LIVING BY CHEMISTRY PROGRAM SAMPLER
©2006 Key Curriculum Press

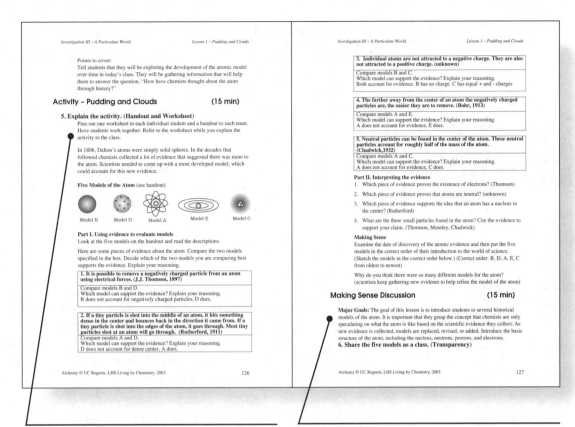

Points to cover:
Tell students that they will be exploring the development of the atomic model over time in today's class. They will be gathering information that will help them to answer the question, "How have chemists thought about the atom through history?"

Activity – Pudding and Clouds   (15 min)

5. Explain the activity. (Handout and Worksheet)
Pass out one worksheet to each individual student and a handout to each team. Have students work together. Refer to the worksheet while you explain the activity to the class.

In 1806, Dalton's atoms were simply solid spheres. In the decades that followed chemists collected a lot of evidence that suggested there was more to the atom. Scientists needed to come up with a more developed model, which could account for this new evidence.

Five Models of the Atom (see handout)

Model B   Model D   Model A   Model E   Model C

Part I. Using evidence to evaluate models
Look at the five models on the handout and read the descriptions.

Here are some pieces of evidence about the atom. Compare the two models specified in the box. Decide which of the two models you are comparing best supports the evidence. Explain your reasoning.

| **1. It is possible to remove a negatively charged particle from an atom using electrical forces. (J.J. Thomson, 1897)** |
| Compare models B and D. Which model can support the evidence? Explain your reasoning. B does not account for negatively charged particles, D does. |

| **2. If a tiny particle is shot into the middle of an atom, it hits something dense in the center and bounces back in the direction it came from. If a tiny particle is shot into the edges of the atom, it goes through. Most tiny particles shot at an atom will go through. (Rutherford, 1911)** |
| Compare models A and D. Which model can support the evidence? Explain your reasoning. D does not account for dense center, A does. |

| **3. Individual atoms are not attracted to a negative charge. They are also not attracted to a positive charge. (unknown)** |
| Compare models B and C. Which model can support the evidence? Explain your reasoning. Both account for evidence. B has no charge. C has equal + and - charges |

| **4. The farther away from the center of an atom the negatively charged particles are, the easier they are to remove. (Bohr, 1913)** |
| Compare models A and E. Which model can support the evidence? Explain your reasoning. A does not account for evidence, E does. |

| **5. Neutral particles can be found in the center of the atom. These neutral particles account for roughly half of the mass of the atom. (Chadwick,1932)** |
| Compare models A and C. Which model can support the evidence? Explain your reasoning. A does not account for evidence, C does. |

Part II. Interpreting the evidence
1. Which piece of evidence proves the existence of electrons? (Thomson)
2. Which piece of evidence proves that atoms are neutral? (unknown)
3. Which piece of evidence supports the idea that an atom has a nucleus in the center? (Rutherford)
4. What are the three small particles found in the atom? Cite the evidence to support your claim. (Thomson, Moseley, Chadwick)

Making Sense
Examine the date of discovery of the atomic evidence and then put the five models in the correct order of their introduction to the world of science. (Sketch the models in the correct order below.) (Correct order: B, D, A, E, C from oldest to newest)

Why do you think there were so many different models for the atom? (scientists keep gathering new evidence to help refine the model of the atom)

Making Sense Discussion   (15 min)

Major Goals: The goal of this lesson is to introduce students to several historical models of the atom. It is important that they grasp the concept that chemists are only speculating on what the atom is like based on the scientific evidence they collect. As new evidence is collected, models are replaced, revised, or added. Introduce the basic structure of the atom, including the nucleus, neutrons, protons, and electrons.
6. Share the five models as a class. (Transparency)

The Activity gives students the chance to investigate some of the questions arising from the ChemCatalyst, and serves to illuminate important concepts from the lesson.

The Making Sense question and discussion ensure that students grasp the meaning of what was carried out in the Activity. In this whole-class discussion, students formulate and share ideas. Suggestions are included for helping students reach conclusions, as are definitions and ideas that are appropriate to share with students at this point.

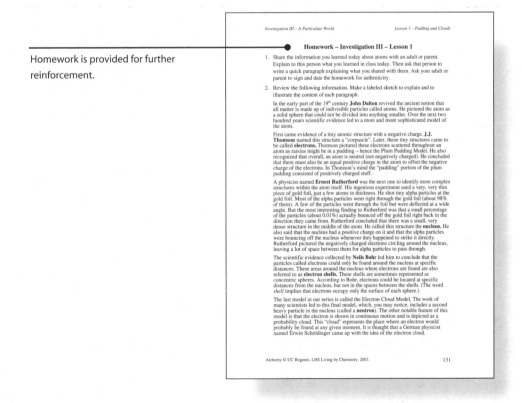

Homework is provided for further reinforcement.

Homework – Investigation III – Lesson 1

1. Share the information you learned today about atoms with an adult or parent. Explain to this person what you learned in class today. Then ask that person to write a quick paragraph explaining what you shared with them. Ask your adult or parent to sign and date the homework for authenticity.

2. Review the following information. Make a labeled sketch to explain and to illustrate the content of each paragraph.

In the early part of the 19th century **John Dalton** revived the ancient notion that all matter is made up of indivisible particles called atoms. He pictured the atom as a solid sphere that could not be divided into anything smaller. Over the next two hundred years scientific evidence led to a more and more sophisticated model of the atom.

First came evidence of a tiny atomic structure with a negative charge. **J.J. Thomson** named this structure a "corpuscle". Later, these tiny structures came to be called **electrons**. Thomson pictured these electrons scattered throughout an atom as raisins might be in a pudding – hence the Plum Pudding Model. He also recognized that overall, an atom is neutral (not negatively charged). He concluded that there must also be an equal positive charge in the atom to offset the negative charge of the electrons. In Thomson's mind the "pudding" portion of the plum pudding consisted of positively charged stuff.

A physicist named **Ernest Rutherford** was the next one to identify more complex structures within the atom itself. His ingenious experiment used a very, very thin piece of gold foil, just a few atoms in thickness. He shot tiny alpha particles at the gold foil. Most of the alpha particles went right through the gold foil (about 98% of them). A few of the particles went through the foil but were deflected at a wide angle. But the most interesting finding to Rutherford was that a small percentage of the particles (about 0.01%) actually bounced off the gold foil right back in the direction they came from. Rutherford concluded that there was a small, very dense structure in the middle of the atom. He called this structure the **nucleus**. He also said that the nucleus had a positive charge on it and that the alpha particles were bouncing off the nucleus whenever they happened to strike it directly. Rutherford pictured the negatively charged electrons circling around the nucleus, leaving a lot of space between them for alpha particles to pass through.

The scientific evidence collected by **Neils Bohr** led him to conclude that the particles called electrons could only be found around the nucleus at specific distances. These areas around the nucleus where electrons are found are also referred to as **electron shells.** These shells are sometimes represented as concentric spheres. According to Bohr, electrons could be located at specific distances from the nucleus, but not in the spaces between the shells. (The word *shell* implies that electrons occupy only the surface of each sphere.)

The last model in our series is called the Electron Cloud Model. The work of many scientists led to this final model, which, you may notice, includes a second heavy particle in the nucleus (called a **neutron**). The other notable feature of this model is that the electron is shown in continuous motion and is depicted as a probability cloud. This "cloud" represents the place where an electron would probably be found at any given moment. It is thought that a German physicist named Erwin Schrödinger came up with the idea of the electron cloud.

LABORATORY KITS

For teachers' convenience, lab kits containing assorted specialty items are available for each unit. The lab kit makes it possible to teach innovative lessons without an inordinate amount of time spent preparing or shopping for unusual items. Kits contain enough materials for several chemistry classes. Refills and replacement parts are also available.

Additional materials needed for teaching each unit, including lab equipment and chemical supplies, are listed in the Teacher Guides (see Materials for This Unit sections in this sampler).

LABORATORY SAFETY

A laboratory setting is ideal for teaching *Living By Chemistry*, but any classroom that meets district safety guidelines can provide a safe setting for these activities. The lab activities can generally be done in a regular classroom if some basic apparatus is available—for instance, squirt bottles and plastic tubs may be used where sinks are not available. Safety is paramount, however, and activities may need to be modified as common sense dictates.

Regardless of the classroom setting, districts should make provisions for safe handling, storage, and disposal of chemicals. Administrators should consult state and district safety guidelines and plan for additional safety training, as needed, for teachers who are either new to the subject or new to the district.

The Units

Each unit consists of approximately 25 lessons designed for a typical 50-minute class period. One unit provides enough material for a 6-week yearlong course. The lessons are clustered into Investigations. Each Investigation addresses a major field of inquiry within the Unit's broader context.

Each section of this sampler showcases a different unit and includes:

- Contents
- Main topics
- Unit overview and investigation summaries
- A complete sample lesson from the Teacher Guide
- The corresponding lesson from the Student Guide
- A list of materials found in the lab kit, and a list of the materials needed to teach the unit

ALCHEMY

Atoms, Elements, and
Compounds

Contents

Main Topics Covered

COVERED IN DEPTH

Periodic table, atomic number, atomic mass

Periodic table, metals, nonmetals, groups

Periodic trends in properties

Periodic table and number of valence electrons

Size and mass of the nucleus

Lanthanide and actinide elements

Electron configurations

Experimental basis for atomic structure

Types of bonds: covalent, metallic, ionic

Ionic bonds

Nuclear forces

Nuclear energy

Isotopes, radioactive isotopes

Radioactive decay

Damage due to radioactive decay

PREREQUISITES FOR THE UNIT

None

Investigation Summaries

The *Alchemy* unit introduces students to the basics of chemistry—atoms, elements, and compounds.

INVESTIGATION I: DEFINING MATTER

The first lessons of *Alchemy* open the door for the study of chemistry by introducing the history of alchemy and establishing basic laboratory safety procedures. Students perform a chemistry experiment in the first lesson, dipping a penny into a liquid and firing it until it appears to be gold. The golden pennies serve as a jumping-off point for learning the definitions of matter, mass, density, and volume.

INVESTIGATION II: BASIC BUILDING MATERIALS

Investigation II introduces students to the elemental building materials of matter, and how scientists relate to those materials. They learn about chemical symbols and formulas, and begin to relate them to their chemical counterparts. They perform a lab experiment (the copper cycle) to demonstrate the immutability of matter.

INVESTIGATION III: A PARTICULATE WORLD

This investigation covers the structure of atoms in detail. Beginning with several atomic models, students discover patterns that relate atomic properties to the modern periodic table. Further lessons include isotopes and electron subshells. A final lab experiment provides exciting evidence for electrons moving between atomic subshells.

INVESTIGATION IV: A SUBATOMIC WORLD

Investigation IV focuses on nuclear chemistry, beginning with an analysis of stable atoms. Moving through alpha and beta decay, students discover that it is possible to transform other elements into gold. They learn that while under normal circumstances this transformation does not occur on Earth, nuclear fusion does take place in supernovas.

INVESTIGATION V: BUILDING WITH MATTER

Investigation V discusses ways in which atoms combine. Students discover that because creating gold is difficult, it may be better to create a substance with properties similar to gold. Students expand their knowledge of atoms as they learn about chemical bonds, the rule of eight, and creating compounds. They are prepared for learning more about molecules in the next unit.

BEFORE CLASS...

LESSON 4 – Create a Table

Key Ideas:

In 1889, Dimitri Mendeleyev, a Russian chemistry teacher, came up with an ingenious organization of the 63 known elements. He organized the elements according to their properties. This organization evolved into our contemporary periodic table. Mendeleev was even able to predict the existence of as-yet undiscovered elements based on gaps located in his table.

What Takes Place:

Students are briefly introduced to Mendeleev and his original idea for organizing the elements into a table. Then they are presented with a deck of thirty-three periodic table cards, which they must sort into a table based on the data on the front of the cards. In essence they are discovering and recreating a portion of the periodic table.

Set-up:

You may wish to cover or remove any periodic tables you have hanging in the room before you complete this activity.

For a demonstration on the reactivity of the elements, put 100 mL of water into two 250 mL beakers. Add a few drops of phenolphthalein to each. Have magnesium and calcium available to put into the beakers.

Materials:

- Student worksheet
- Create a Table cards – 8 sets of 33 cards*
- Transparencies – ChemCatalyst and Check-in
- 1-2 g magnesium
- 1-2 g calcium
- 2 250-mL beakers
- a few drops phenolphthalein
- water

---

*Black-and-white versions of these cards are printed in this sampler on pages 172–178. To try this activity, make single-sided copies of the cards, preferably onto card stock, and cut them out. The cards are included in the *Alchemy* lab kit and may also be purchased separately from Key Curriculum Press at (800) 995-6284.

Investigation II – Basic Building Materials
LESSON 4 – Create a Table

Students are told about Mendeleyev's efforts to organize the elements into a table, and are asked to perform a similar task using cards corresponding to the first 33 elements (omitting germanium). The cards contain information related to the properties of the elements. The students essentially "discover" the basic organization of the periodic table of the elements based on the properties and data found on the cards. (Note: Before you begin this class you may wish to remove or cover any periodic tables you may have hanging in the room.)

Exploring the Topic   (10 min)

1. Introduce the ChemCatalyst exercise. (Transparency)

Display the transparency with the following exercise for students to complete individually.

In 1889 a Russian chemistry teacher created an organized table of the elements. At the time only 63 different elements were known. Below is a reproduction of that table.

Mendeleyev's Table of the Elements – 1889

	Group I	Group II	Group III	Group IV	Group V	Group VI	Group VII	Group VIII
1	H = 1							
2	Li = 7	Be = 9	B = 11	C = 12	N = 14	O = 16	F = 19	
3	Na = 23	Mg = 24	Al = 27	Si = 28	P = 31	S = 32	Cl = 35	
4	K = 39	Ca = 40	___ = 44	Ti = 48	V = 51	Cr = 52	Mn = 55	Fe = 56 Co = 59 Ni = 59 Cu = 63
5	Cu = 63	Zn = 65	___ = 68	___ = 72	As = 75	Se = 78	Br = 80	
6	Rb = 85	Sr = 87	Yt = 88	Zr = 90	Nb = 94	Mo = 96	___ = 100	Ru =104 Rh =106 Pd =106 Ag =108
7	Ag = 108	Cd = 112	In = 113	Sn = 118	Sb = 122	Te = 125	I = 127	
8	Cs = 133	Ba = 137	Di = 138	Ce = 140	___	___		___
9	___	___						
10	___		Er = 178	La = 180	Ta = 182	W = 184	___	Os = 195 Ir =197 Pt =198 Au = 199
11	Au = 199	Hg = 200	Tl = 204	Pb = 207	Bi = 208		___	___
12		___	___	Th = 231	___	U = 240	___	___

Note: Mendeleyev's symbol for iodine, "J", has been changed to "I" to match modern symbols.

- How do you think the elements are organized?
- What do you think the numbers represent?

Alchemy © UC Regents, LHS Living by Chemistry, 2003.   89

2. Discuss the ChemCatalyst exercise.

If you wish to give credit for the ChemCatalyst question, collect student answers either before or after the discussion.

Discussion goals:
Use the historical context to introduce the periodic table. Explore students' existing knowledge on the subject.

Sample questions:
How do you think the elements are organized in this document?
What do you think the numbers represent?
What do you notice about the numbers when you go across the table?
  Down? (numbers increase going across and down the table)
There are several blank spaces in the table – what do you think these
  represent?

This is an open-ended discussion. The students may or may not have heard of the periodic table. This is all right. Some may even know about Mendeleyev's work. They may identify the numbers as having something to do with the weight or the mass of each element. Or, students may use the terms atomic mass or atomic weight. It is not necessary to identify these terms yet.

3. Introduce Mendeleyev.

Share all or part of the following information with the class about Mendeleyev. Write the symbols for the four elements, beryllium, magnesium, calcium, and strontium on the board in a column as shown below.

Some information to share:
The document in the ChemCatalyst is a reproduction of a table created by **Dimitri Mendeleyev** in 1889. Mendeleyev is credited with organizing the elements into the first **periodic table**. He was born in Siberia in 1834, the 17th child in a very large family. His father died when he was 13 years of age and his mother decided to move what was left of her family to the city to get her youngest son into a university. He was turned down for admission to Moscow University so she pressed on to St. Petersburg. He was turned down there for medical school, mostly because of his humble origins. His mother died just 10 weeks after he was finally accepted into a teaching institute in St Petersburg.

Ultimately, Mendeleyev became a chemistry professor and became interested in organizing the elements. Mendeleyev's organization concentrated on the properties of the elements, specifically how they reacted with one another. It is said that he went to sleep one night and dreamt of a table where the elements were organized by similar properties. In his mind they simply fell into their appropriate places. Upon awakening he applied his dream to the task at hand, finding patterns in the repetitive properties of the elements. Mendeleyev created his first table of the elements in

| Be |
| Mg |
| Ca |
| Sr |

Alchemy © UC Regents, LHS Living by Chemistry, 2003.   90

1869. The table shown in the ChemCatalyst represents a later version created by Mendeleyev in 1889. On the board is an example of a group of elements that Mendeleyev placed together. The names of these elements are beryllium, magnesium, calcium, and strontium.

The main properties that Mendeleyev used to sort the elements were reactivity (described below), and a number describing the **atomic weight** of each element. Back in Mendeleyev's day, chemists were able to calculate this number by looking at the proportions in which elements combined with each other to form new substances.

Students do not need to know how atomic weight was arrived at back in Mendeleyev's time. Nevertheless, a simplified explanation follows: The element of copper always combines in the same proportion with carbon and oxygen to form copper carbonate, $CuCO_3$. Every sample of copper carbonate that is examined will always be 51.5% copper, 38.8% oxygen, and 9.7% carbon, or a ratio of 5.3 : 4 : 1. Chemists were able to break copper carbonate down into its constituent elements and find the mass of each portion. From that they calculated the percentage and proportion of each portion. If you look in Mendeleyev's table you will find that the numbers corresponding to the atomic weights of these elements, are in this same proportion.

4. Demonstrate the concept of reactivity.

Suggested demonstration:
1. Place approximately 100 mL of water in each of two 250 mL beakers
2. Add a few drops of phenolphthalein to the water in each beaker.
3. Place a piece of white paper behind both beakers to improve visibility for the class.
4. Add a small piece of calcium (1-2 g) to one beaker and a small piece of magnesium (1-2 g) to the other.
5. Observe what happens.

Sample questions:
  What do you observe?
  Which element is more reactive? How do you know?

Points to cover:
Tell students that you are going to demonstrate the **reactivity** of two of the elements that Mendeleyev grouped together. In the beaker with the calcium you should observe the appearance of a bright pink color and some bubbling. The beaker with the magnesium will not appear to be doing much of anything. Within about twenty minutes, however, the magnesium will also show evidence of reacting with the water. Both elements are similar in that they both react with water. However, the calcium is considered **more reactive** than the magnesium because of the speed at which it reacts when it comes into contact with water. If we look at the next element listed in Mendeleev's grouping, strontium, Sr, we find an even more reactive element. Strontium reacts even more vigorously with water than calcium.

Alchemy © UC Regents, LHS Living by Chemistry, 2003.   91

5. Explain the purpose of today's activity.

If you wish you can write the main question on the board.

Points to cover:
Tell students they will be given a set of cards that have information on them associated with thirty-three different elements. They will try to come up with the remainder of the periodic table using the grouping given above as a starting point. The goal is to answer the following question, "How did Mendeleyev organize the elements?"

Activity – Create a Table   (15–20 min)

6. Introduce the activity.

Explain the procedure to the class. Pass out sets of cards to teams of students. (Sample card shown below.)

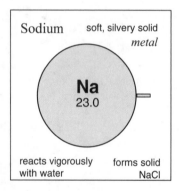

Instructions:
1. Work in teams with one set of Create a Table cards.
2. Find Be, Mg, Ca, and Sr from the deck of cards, and arrange them in a column like Mendeleyev did.
3. With your team, decide how you can use the remainder of the cards to organize the elements into a table similar to Mendeleyev's.
4. Arrange the cards and answer the following questions.

Questions on the worksheet:
1. What characteristics did you use for sorting the cards?
2. Where did you put H and He? What was your reasoning for their placements?
3. Did you notice any cards that didn't quite fit, or that seemed out of order? Explain.

Alchemy © UC Regents, LHS Living by Chemistry, 2003.   92

Making Sense:

Below are five possible cards for the element germanium. Where does germanium belong in the table? Which card seems most accurate to you? What is your reasoning?

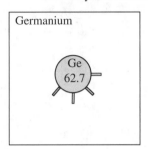

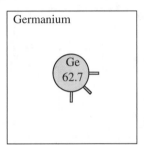

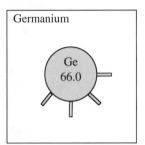

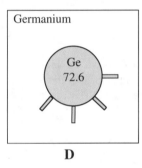

D

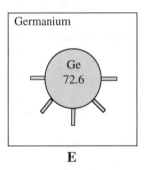

E

What would you add to the three empty corners to complete the card?

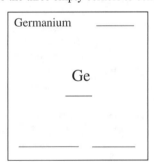

Circulate amongst the students to monitor and assist them. They should be reconstructing Mendeleyev's table around the existing column). You can inform them that Mendeleyev found gaps in his periodic table and predicted that new elements would be found to fill those gaps. If students are having trouble with H and He, have them save these two elements for last.

Completed table

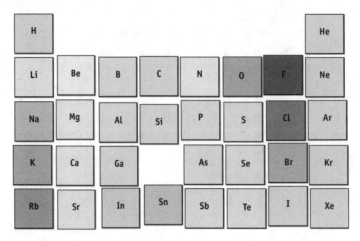

Making Sense Discussion   (10–15 min)

Major Goals: The main goal of this discussion is to provide an opportunity for teams to share their sorting experience and to begin to identify the many patterns they discovered while they were organizing their cards. These patterns will be dealt with in more depth in the following lesson.

7. Discuss the sorting activity.

<u>Discussion goals:</u>
Allow students to share their experience of organizing the cards, as well as some of their discoveries.

Sample questions:
  Describe the process you went through in your teams to sort the cards.
  What characteristics did you use for sorting?
  Did you have to abandon any characteristics and pick different ones?
  Did you discover any new characteristics or patterns during the activity?

Many groups will use color and size of the circles as characteristics to sort with. Others will notice the atomic weight in the middle of each card. Still others will notice that the number of "spokes" on the outside of each circle varies in a systematic way. It is not necessary to define these characteristics formally as yet. Some groups may have their table sorted from right to left, rather than left to right. Others may have their table organized vertically rather than horizontally. These are all fine.

8. Discuss the missing element.
As students give you information, construct a card for germanium on the board.

Discussion goals:
Ascertain how students constructed a card for the missing element, germanium.

Sample questions:
  Which card did you choose for Ge? Explain your choice.
  Do you predict that Ge will be a solid, liquid, or gas?
  Do you predict that Ge will be shiny? Explain.
  How do you expect Ge will react with air? Explain.
  Do you expect Ge to form GeH_2, GeH_3, or GeH_4?
  How you think Mendeleyev was able to predict new elements before they were discovered? (He used the periodic table and trends in properties to identify missing elements such as germanium.)

Card D is the most accurate one. The number on this card is somewhere between the numbers of gallium and arsenic, which are on either side of it in the table on the transparency. The size of the circle shown fits best between gallium and arsenic horizontally, and silicon and tin vertically. Also, the element should have four spokes. Regarding the three missing corners of the card, students should say something close to the following: 1) moderately soft, silvery solid, metalloid; 2) forms GeH_4 gas; and 3) corrodes very slowly in air.

Check-in   (5 min)

9. Introduce the Check-in exercise.

Write the following Check-in question on the board. Students should answer the question individually.

  • Which of the following elements would you find in the same group on the periodic table? Explain your thinking.

Cadmium Cd	Moderately soft, silvery solid, metal	React very slowly with water	Found in $CdCl_2$ (s)
Zinc Zn	Moderately hard, silvery solid, metal	Reacts very slowly with water	Found in $ZnCl_2$ (s)
Iodine I	Purple solid, nonmetal	Reacts slowly with metals	Found in ICl (s)
Mercury Hg	Silvery liquid, metal	Does not react with water	Found in $HgCl_2$ (s)

10. Discuss the Check-in exercise.

Get a sense of the level of understanding by taking a vote, collecting students' work, or asking students to defend their choices.

Discussion goals:
Ascertain whether students can see and describe the general trends in the periodic table.

Alchemy © UC Regents, LHS Living by Chemistry, 2003.   95

Sample questions:
> Which elements should be grouped together? Explain your choice.
> Which of the elements that you group together would you put at the top of the column? at the bottom? Explain.
> What other information would you want to know to check your choice?

Zinc, cadmium, and mercury should be grouped together. They are all metals. They get softer from zinc to cadmium, and mercury is a liquid. They are all not very reactive and all are found in compounds with the chemical formula MCl_2. We have observed that metals get softer as you go down a group. Thus, we expect the order of the elements to be zinc, cadmium, and mercury from the top of the group to the bottom. If would be good to know the atomic weights to verify this. Iodine is very different from these three elements.

11. Wrap-up

Assist the students in summarizing what was learned in the class.

- Mendeleyev organized the periodic table based on the properties of the elements.
- Mendeleyev's arrangement of the elements helped to predict the existence of undiscovered elements.

Homework

12. Assign homework.

Use the homework provided with the curriculum or assign your own.

TEACHER GUIDE

Homework – Investigation II – Lesson 4

1. Using the Encyclopedia, the Internet or some other reference source, do some research on Dimitri Mendeleyev, the Russian chemist credited with the discovery of the periodic table of the elements.

    a) What properties or characteristics did Mendeleyev use to sort the elements in his periodic table?

    b) Write a brief paragraph describing Mendeleyev's life and work.

2. Using the Encyclopedia, the Internet or some other reference source, find illustrations of the periodic table.

    a) Print or copy at least two different versions of the periodic table and bring them in to class.

    b) Write down what you think makes your two versions different from each other.

ChemCatalyst: This document was created in 1889, when chemists only knew of 63 different elements. How do you think the elements are organized? What do you think the numbers represent?

Transparency

98

Mendeleyev's Table of the Elements - 1889

	Group I	Group II	Group III	Group IV	Group V	Group VI	Group VII	Group VIII
1	H = 1							
2	Li = 7	Be = 9	B = 11	C = 12	N = 14	O = 16	F = 19	
3	Na = 23	Mg = 24	Al = 27	Si = 28	P = 31	S = 32	Cl = 35	
4	K = 39	Ca = 40	— = 44	Ti = 48	V = 51	Cr = 52	Mn = 55	Fe = 56 Co = 59 Ni = 59 Cu = 63
5	Cu = 63	Zn = 65	— = 68	— = 72	As = 75	Se = 78	Br = 80	
6	Rb = 85	Sr = 87	Yt = 88	Zr = 90	Nb = 94	Mo = 96	= 100	Ru =104 Rh =106 Pd =106 Ag =108
7	Ag = 108	Cd = 112	In = 113	Sn = 118	Sb = 122	Te = 125	I = 127	
8	Cs = 133	Ba = 137	Di = 138	Ce = 140	—	—	—	
9	—							—
10			Er = 178	La = 180	Ta = 182	W = 184	—	Os = 195 Ir =197 Pt =198 Au =199
11	Au = 199	Hg = 200	Tl = 204	Pb = 207	Bi = 208	—	—	—
12	—			Th = 231	—	U = 240		

Note: Mendeleyev's symbol for iodine, "J", has been changed to "I" to match modern symbols.

Alchemy © UC Regents, LHS Living by Chemistry, 2003.

Check-in: Which of the following elements would find in the same group on the periodic table? Explain your thinking.

Transparency

Cadmium Cd	Moderately soft, silvery solid, metal	Reacts very slowly with water	Found in $CdCl_2$ (s)
Zinc Zn	Moderately hard, silvery solid, metal	Reacts very slowly with water	Found in $ZnCl_2$ (s)
Iodine I	Purple solid, nonmetal	Reacts slowly with metals	Found in ICl (s)
Mercury Hg	Silvery liquid, metal	Does not react with water	Found in $HgCl_2$ (s)

99

Alchemy © UC Regents, LHS Living by Chemistry, 2003.

Create a Table

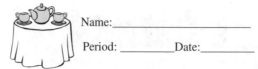

Name:_____

Period: _____Date:_____

Purpose: The goal of this lesson is to acquaint you with Mendeleyev's organization of the elements by allowing you to create your own table from the patterns you see in the elements.

Instructions:
1. Work in teams with one set of periodic table cards.
2. Find Be, Mg, Ca, and Sr from the deck of cards , and arrange them in a column like Mendeleyev did.
3. With your team, decide how you can use the remainder of the cards to organize the elements into a table similar to Mendeleyev's.
4. Arrange the cards and answer the following questions.

| Be |
| Mg |
| Ca |
| Sr |

Answer the following questions:
1. What characteristics did you use for sorting the cards?

2. Where did you put H and He? What was your reasoning for their placements?

3. Did you notice any cards that didn't quite fit, or that seemed out of order? Explain.

TEACHER GUIDE

Making Sense:

Below are five possible cards for the element germanium. Where does germanium belong in the table? Which card seems most accurate to you? What is your reasoning?

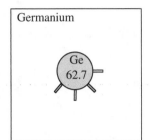

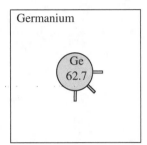

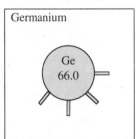

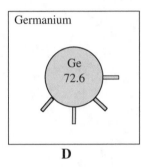

D

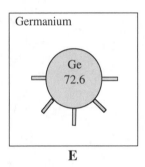

E

What would you add to the three empty corners to complete the card?

Germanium	_____
	Ge
	___
_____	_____
	___

Lesson Guide: Create a Table

Investigation II – Lesson 4

The goal of this lesson is to acquaint you with an organization of the elements called the periodic table. This activity allows you to create your own table from a set of cards using the patterns you see in the properties of the elements.

ChemCatalyst

Mendeleyev, a Russian chemistry teacher, created an organized table of the elements in 1889. At the time only 63 different elements were known. Below is a reproduction of that table.

Mendeleyev's Table of the Elements - 1889

	Group I	Group II	Group III	Group IV	Group V	Group VI	Group VII	Group VIII
1	H = 1							
2	Li = 7	Be = 9	B = 11	C = 12	N = 14	O = 16	F = 19	
3	Na = 23	Mg = 24	Al = 27	Si = 28	P = 31	S = 32	Cl = 35	
4	K = 39	Ca = 40	___ = 44	Ti = 48	V = 51	Cr = 52	Mn = 55	Fe = 56 Co = 59 Ni = 59 Cu = 63
5	Cu = 63	Zn = 65	___ = 68	___ = 72	As = 75	Se = 78	Br = 80	
6	Rb = 85	Sr = 87	Yt = 88	Zr = 90	Nb = 94	Mo = 96	___ = 100	Ru =104 Rh =106 Pd =106 Ag =108
7	Ag = 108	Cd = 112	In = 113	Sn = 118	Sb = 122	Te = 125	I = 127	
8	Cs = 133	Ba = 137	Di = 138	Ce = 140	___	___	___	___
9	___	___	___		___	___	___	
10	___		Er = 178	La = 180	Ta = 182	W = 184	___	Os = 195 Ir =197 Pt =198 Au = 199
11	Au = 199	Hg = 200	Tl = 204	Pb = 207	Bi = 208		___	___
12	___	___		Th = 231	___	U = 240	___	___

Note: Mendeleyev's symbol for iodine, "J", has been changed to "I" to match modern symbols.

- How do you think the elements are organized?
- What do you think the numbers represent?

One of the properties you will be considering today is called **reactivity.** Different elements may react similarly when they come in contact with the same substance. Thus, calcium and magnesium both react with water. However, calcium is considered **more reactive** than magnesium because of the *speed* at which it reacts once it comes into contact with water.

Activity

Instructions:

1. Work in teams of four with one set of periodic table cards.

2. Find Be, Mg, Ca, and Sr from the deck of cards, and arrange them in a column like Mendeleyev did.

3. With your team, decide how you can use the remainder of the cards to organize the elements into a table similar to Mendeleyev's.

4. Arrange the cards and answer the following questions.

Answer the following questions:

1. What characteristics did you use for sorting the cards?

2. Where did you put H and He? What was your reasoning for their placements?

3. Did you notice any cards that didn't quite fit, or that seemed out of order? Explain.

Making Sense Question:

Below are five possible cards for the element germanium. Where does germanium belong in the table? Which card seems most accurate to you? What is your reasoning?

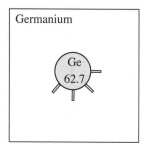

A

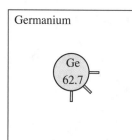

B

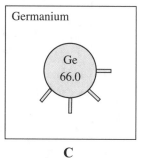

C

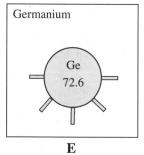

D

E

Alchemy © UC Regents, LHS Living by Chemistry, 2003

LIVING BY CHEMISTRY PROGRAM SAMPLER
©2006 Key Curriculum Press

STUDENT GUIDE

Draw a card similar to the one below in your notebook. What would you add to the three empty corners to complete the card?

Germanium _____
Ge
_____
_____   _____

Check-in

Answer the following question:

- Which of the following elements would you find in the same group on the periodic table? Explain your thinking.

Cadmium Cd	Moderately soft, silvery solid, metal	React very slowly with water	Found in $CdCl_2$ (s)
Zinc Zn	Moderately hard, silvery solid, metal	Reacts very slowly with water	Found in $ZnCl_2$ (s)
Iodine I	Purple solid, nonmetal	Reacts slowly with metals	Found in ICl(s)
Mercury Hg	Silvery liquid, metal	Does not react with water	Found in $HgCl_2$ (s)

Homework

Complete the following for homework:

1. Using the Encyclopedia, the Internet, or some other reference source, do some research on Dimitri Mendeleyev, the Russian chemist credited with the discovery of the periodic table of the elements.

   a) What properties or characteristics did Mendeleyev use to sort the elements in his periodic table?

   b) Write a brief paragraph describing Mendeleyev's life and work.

2. Using the Encyclopedia, the Internet or some other reference source, find illustrations of the periodic table.

   a) Print or copy at least two different versions of the periodic table and bring them in to class.

   b) Write down what you think makes your two versions different from each other.

Materials for *Alchemy* Unit

KIT CONTENTS

8 sets of Create a Table cards (33 cards per set)

8 decks Salty Eights cards

8 Periodic Table Card Sort posters

4 oz sand[R]

8 pieces paraffin wax (0.5 in. $\times$ 2.5 in. $\times$ 5 in.)[R]

65 ft nichrome wire, 20 gauge[R]

50 ft copper wire, bare, 20 gauge[R]

1 wire cutter

8 2-in. brass rods (0.5 in. diameter)

8 2-in. aluminum rods
(0.5 in. diameter)

8 4-in. aluminum rods
(0.5 in. diameter)

96 4-dram glass vials, screw top

4 sets of 18 labels with
vial number, name, formula

8 small lightbulbs

8 bases for bulbs

8 9-volt batteries[R]

12 snap connectors
for 9-volt batteries

30 plastic-coated wires with alligator clips

For up-to-date information about kit contents, please visit www.keypress.com/chemistry .

[R]Consumable; will need to be replenished. This kit includes enough for approximately 10 classes of 32 students, or 2–3 years.

OTHER MATERIALS NEEDED

Common lab ware, supplies, chemicals, and other easily obtainable materials have not been included in this kit. They are listed below.

Your classroom should have a sink. Please note that Bunsen burners can be substituted for hot plates, but hot plates are safer to use.

Teachers should follow all state, local, and district guidelines concerning laboratory safety practices, safe handling of materials, and disposal of hazardous materials. In the absence of guidelines related to the handling, storage, and disposal of laboratory materials, teachers should refer to the Materials Safety Data Sheets (MSDS) for each chemical.

Lab Ware

(These quantities are for a class of 32 students.)

32 pairs of safety goggles

8 Bunsen burners

8 hot plates

18 100-mL beakers

8 250-mL beakers

8 scales (electric preferred)

16 sets of tongs

8 50-mL graduated cylinders

8 small graduated cylinders (or pipettes)

8 funnels and filter paper

8 stirring rods

2–4 magnetic stirrers

16 paper clips

Chemicals

(These quantities are for ten classes of 32 students.)

20 g copper nitrate (cupric nitrate, trihydrate)

30 g copper sulfate (cupric sulfate, pentahydrate)

15 g copper hydroxide (cupric hydroxide)

15 g copper (II) oxide (cupric oxide)

20 g zinc sulfate, heptahydrate, granular

15 g copper, 150 mesh, granular

200 g zinc, 20 mesh, granular

30 g magnesium ribbon

10 g calcium turnings

15 g strontium nitrate

15 g strontium chloride, hexahydrate

15 g potassium nitrate, crystal

15 g potassium sulfate, granular

15 g potassium chloride, granular

15 g sodium carbonate

15 g sodium nitrate

15 g sodium chloride

15 g calcium chloride, dihydrate, flake

15 g sodium hydroxide pellets

LESSON-BY-LESSON MATERIALS GUIDE

The kit contains the specialized items needed for each lab. However, common lab ware, chemicals, and other easily obtainable materials have not been included in this kit. Kit materials and other needed materials are listed separately below.

For consumable items, one kit contains enough materials for 10 classes of 32 students.

Lesson I-1: Penny for Your Thoughts

Materials Provided in Kit	Other Materials Needed
none	32 pairs of safety goggles
	8 copper pennies or more
	~150 g zinc filings
	200 mL of 3.0 M sodium hydroxide (25 mL per team)
	16 sets of tongs
	16 100-mL beakers
	8 250-mL beakers
	8 hot plates
	8 Bunsen burners
	tap water

Lesson I-4: All That Glitters

Materials Provided in Kit	Other Materials Needed
8 2-in. brass rods (0.5-in. diameter)	8 50-mL graduated cylinders
8 2-in. aluminum rods (0.5-in. diameter)	8 scales (1 per team if possible)
8 4-in. aluminum rods (0.5-in. diameter)	~200 mL tap water (25 mL per team)
	8 golden pennies from Lesson I-1
	1 Sacagawea Golden Dollar (optional)

Lesson II-1: A New Language

Preparation: Before class, place the labels on the glass vials and put the appropriate chemicals into the vials. Students will not be opening the vials, so in the case of gases, you may leave the vials empty. For aqueous solutions, use plain water.

Materials Provided in Kit	Other Materials Needed
80 4-dram glass vials, screw top	5 g zinc
4 sets of labels for vials (see table below)	10 g copper nitrate
	10 g copper sulfate
	5 g copper hydroxide
	5 g copper (II) oxide
	10 g zinc sulfate
	10 g sodium hydroxide
	10 g sodium nitrate

Here is the information found on each set of labels for Vials 1–18.

	Chemical Name	Chemical Formula
Vial 1	sodium nitrate	$NaNO_3$ (aq)
Vial 2	copper nitrate	$Cu(NO_3)_2$ (s)
Vial 3	copper hydroxide	$Cu(OH)_2$ (s)
Vial 4	hydrogen	H_2 (g)
Vial 5	sodium nitrate	$NaNO_3$ (s)
Vial 6	sodium hydroxide	NaOH (s)
Vial 7	copper sulfate	$CuSO_4$ (s)
Vial 8	zinc	Zn (s)
Vial 9	nitric acid	HNO_3 (aq)
Vial 10	copper	Cu (s)

	Chemical Name	Chemical Formula
Vial 11	sodium hydroxide	NaOH (aq)
Vial 12	copper oxide	CuO (s)
Vial 13	water	H_2O (l)
Vial 14	sulfuric acid	H_2SO_4 (aq)
Vial 15	zinc sulfate	$ZnSO_4$ (aq)
Vial 16	copper sulfate	$CuSO_4$ (aq)
Vial 17	copper nitrate	$Cu(NO_3)_2$ (aq)
Vial 18	copper sulfate and copper	$CuSO_4$ (aq) and Cu (s)

Disposal: Be sure to dispose of chemicals properly. Check state and district guidelines, or refer to the Materials Safety Data Sheet for each chemical. (You may store the filled, labeled vials for reuse, but over time some of the solids may change in appearance.)

Lesson II-2: Now You See It . . .

Materials Provided in Kit	Other Materials Needed
none	8 g copper powder (1 g per team)
	8 g zinc (1 g per team)
	160 mL 8M nitric acid (20 mL per team)
	1200 mL 1M sulfuric acid (150 mL per team)
	160 mL 8M sodium hydroxide (20 mL per team)
	8 250-mL beakers
	8 100-mL beakers
	8 graduated cylinders (or pipettes)
	2–4 sets of tongs
	8 funnels and filter paper
	8 stirring rods
	2–4 hot plates
	2–4 magnetic stirrers

Disposal: Be sure to dispose of chemicals properly. Check state and district guidelines, or refer to the Materials Safety Data Sheet for each chemical.

Lesson II-4: Create a Table

Materials Provided in Kit	Other Materials Needed
8 sets of Create a Table cards (33 cards per set)	2 g magnesium
	2 g calcium
	a few drops of phenolphthalein indicator
	2 250-mL beakers
	water

Lesson II-5: Breaking the Code

Materials Provided in Kit	Other Materials Needed
8 Periodic Table Card Sort posters	none

Lesson III-7: Technicolor Atoms

Preparation: Before the lab, cut enough pieces of nichrome and copper wire. Make 0.5 M solutions from the chemicals listed in the right-hand column below.

Materials Provided in Kit	Other Materials Needed
22 6-in. pieces of nichrome wire	8 sets of tongs
8 6-in. pieces of copper wire	8 copper pennies
1 wire cutter	8 Bunsen burners (1 per team)
	10 g strontium nitrate
	10 g strontium chloride
	10 g copper sulfate
	10 g copper nitrate
	10 g sodium carbonate
	10 g sodium nitrate
	10 g sodium chloride
	10 g potassium nitrate
	10 g potassium sulfate
	10 g potassium chloride

Disposal: Be sure to dispose of chemicals properly. Check state and district guidelines, or refer to the Materials Safety Data Sheet for each chemical.

Lesson V-1: You Light Up My Life

Note: In place of bulbs and bases, you may also use bulbs cut from a string of holiday lights.

Materials Provided in Kit	Other Materials Needed
8 bulbs (1 per team)	16 paper clips (2 per team)
8 bases for bulbs	1 packet Kool-Aid®
8 battery snap connectors	18 100-mL beakers (2 per team)
24 alligator clips (3 per team)	distilled water
8 9-volt batteries	aluminum foil
8 pieces paraffin wax (1 per team)	sugar
4 oz sand (0.5 oz per team)	copper
	calcium chloride
	copper sulfate
	salt
	ethyl alcohol

Lesson V-1: You Light Up My Life *(continued)*

Preparation: Assemble the conductivity apparatus as shown.

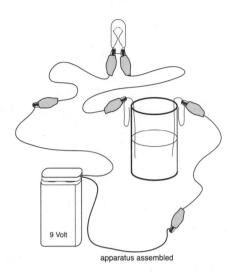

apparatus assembled

Lesson V-5: Salty Eights

Materials Provided in Kit	Other Materials Needed
8 decks of Salty Eights cards	none

SMELLS
Molecular Structure and Properties

Contents

Main Topics Covered

COVERED IN DEPTH

Periodic trends—electronegativity

Periodic table—number of valence electrons

Covalent bonds

Intermolecular forces

Lewis dot structures

Shape and polarity

Random motion of gases

Bonding characteristics of carbon

Functional groups

INTRODUCED OR COVERED BRIEFLY

Ionic bonds

Chemical reactions

Role of a catalyst

Large molecules (polymers)

PREREQUISITES FOR UNIT

Periodic table—elemental symbols and atomic numbers

Periodic trends—introduction to the number of valence electrons

Types of bonds—covalent vs. ionic

LIVING BY CHEMISTRY PROGRAM SAMPLER
©2006 Key Curriculum Press

Investigation Summaries

This unit is a comprehensive study of molecular structure and properties, beginning with chemical formulas and ending with three-dimensional models. Students explore the relationship between scent and chemistry, and cover topics such as functional groups, bonding, and the forces that dictate molecular behavior.

INVESTIGATION I: SPEAKING OF MOLECULES

Investigation I uses students' experiences with scent to create concrete connections between smell and chemistry. Using strongly scented substances, the class investigates the important, but not entirely consistent, relationship between scent and molecular formula.

INVESTIGATION II: PICTURING MOLECULES

Investigation II moves students from an intuitive knowledge of molecular structure to a concrete one by introducing molecular diagrams of various types. By investigating molecular shape, students also learn about functional groups and their effect on scent. A lab activity allows students to put that knowledge to use, creating molecules with specific scents.

INVESTIGATION III: BUILDING MOLECULES

Now that students have grasped molecular structure in two dimensions, Investigation III moves their thinking from a plane into three-dimensional space, using models to increase student understanding of molecular structure. This investigation emphasizes three-dimensional structure and addresses lone and bonded pairs of electrons, basic bonding, and isomers.

INVESTIGATION IV: MOLECULES IN ACTION

Investigation IV introduces students to the process by which we smell certain substances and not others, including the receptor-site model and basic physiology. It also addresses the forces that affect molecules and their interactions, such as bonding, electronegativity, and dipoles.

BEFORE CLASS . . .
LESSON 4 – Polar Bears and Penguins

Key Ideas:
The concept of **electronegativity** is introduced as a measure of the tendency of an atom to attract a shared pair of electrons when it is part of a molecule. When two different atoms bond together, they do not attract the shared electron pair between them equally. If one atom has a greater tendency to attract the shared electrons, then it will have a partial negative charge relative to the other atom, which will have a partial positive charge. There is a continuum of bonding from equal sharing to unequal sharing due to large differences in electronegativity.

What Takes Place:
Students read a comic strip about polarity, electronegativity, and bonding. They answer questions on a worksheet aimed at assisting them in analyzing the comic strip. The concept of electronegativity is introduced to the class along with three different categories of bonds.

Materials:
- Student worksheet
- Polarity comic strip – one for each student
- Transparency – ChemCatalyst

TEACHER GUIDE

Investigation IV – Molecules in Action
LESSON 4 – Polar Bears and Penguins

Students explore polarity further by interpreting the scenes from a comic book on the topic in which penguins and polar pairs are bonded to one another through an ice cream cone with scoops of ice cream. The polar bears win out over the penguins in attracting the scoops to their mouths. This is an analogy for electron pairs in bonds, which tend to be attracted towards one of the atoms sharing the electrons. Electronegativity is introduced as a measure of this tendency and as an explanation for partial charges on atoms in molecules, which give rise to polarity.

Exploring the Topic   (10 min)

1. Introduce the ChemCatalyst exercise. (Transparency)
Display the following ChemCatalyst question for students to complete individually.

Consider the following illustration:

- Draw the Lewis dot structure for HCl.

- If the penguin represents a hydrogen atom and the polar bear represents a chlorine atom, what does the ice cream represent in the drawing? What do you think the picture is trying to illustrate?

- Would HCl be attracted to the charge wand? Explain your thinking.

2. Discuss the ChemCatalyst exercise.
Use the discussion to get a sense of the students' initial ideas.

248

LIVING BY CHEMISTRY PROGRAM SAMPLER
©2006 Key Curriculum Press

<u>Discussion goals:</u>
Begin to discuss unequal sharing of electrons.

Sample questions:

What do you think the drawing is trying to illustrate?

Why is the larger animal representing the chlorine atom?

Why is the penguin being swept off his feet by the polar bear?

Do you think HCl will be attracted to a charged wand?

Students should have a wide range of individualized responses to these questions. We hope they will deduce that the ice cream scoops represent bonded electrons. The polar bear appears to be pulling on the bonded electrons more strongly than the penguin. The polar bear is a bigger animal representing a bigger atom. The chlorine atom pulls on the bonded pair of electrons harder than the hydrogen atom. This means that the electrons are not shared evenly. Some students may begin to speculate that this would cause a partial charge and that HCl would therefore be attracted to a charged wand. Simply accept/acknowledge all ideas at this point.

3. Refine definition of polarity.

Provide students with a simple explanation of the ChemCatalyst cartoon as a bridge to the activity. Keep the ChemCatalyst transparency on the board for this discussion.

<u>Discussion goals:</u>
Begin to discuss how partial charges are created. Use the discussion as a springboard into the comic book activity.

Sample questions:

What did the charged wand demonstrate in the last activity? (the presence of a charge on the molecules of certain liquids)

In what ways do polar molecules differ from nonpolar molecules? (intermolecular interactions, boiling point, etc.)

What explanation for partial charges does the ChemCatalyst illustration provide? (see paragraph below)

Points to cover:
In the last lesson it was demonstrated that certain liquids are attracted to a charged wand because of partial charges on the molecules. However, we did not have any explanation for the existence of these partial charges. The polar bear-penguin cartoon provides a symbolic explanation for polarity. The chlorine atom, represented by the polar bear, attracts the pair of bonded electrons that are shared between the hydrogen atom and the chlorine atom more strongly. In fact, the chlorine atom attracts these shared electrons so strongly that they are more often found around the chlorine atom than the hydrogen atom. This results in a partial negative charge on the chlorine atom and a partial positive charge on the hydrogen atom. Today's activity will explore this tendency to attract a shared pair of electrons in more detail.

Smells © UC Regents, LHS Living by Chemistry, 2003.   249

TEACHER GUIDE

4. Explain the purpose of the activity

If you wish you can write the main question on the board.

Points to cover:
Tell students they will be reading a comic book on the subject of polarity. They will be gathering information that will assist them in answering the question: "How can we explain partial charges on molecules?"

Activity – Polar Bears and Penguins   (15 min)

5. Analyze the polarity comic book. (Comic Book and Worksheet)

Pass out the comic book called "The Bare Essentials of Polarity". Have the students read the comic strip through once before passing out the worksheet. Then ask them to work in pairs to answer the following questions, which are on the worksheet.

Use the comic book called "The Bare Essentials of Polarity" to answer the following questions.

1. How does the comic book define a "polar molecule"?
2. Use your own words to define electronegativity as you understand it, after reading the first two pages of the comic book.
3. Interpret the picture at the bottom of page 1. Explain how the iceberg, penguins, and polar bears represent trends in electronegativity.
4. What is the artist trying to represent when there are two polar bears arm wrestling together, or two penguins arm wrestling together?
5. What three types of bonds are represented on the third page of the comic book? What happens to the bonding electrons in each type of bond?
6. Explain why there are four scoops of ice cream in the illustration of O_2 on page 3. (There are four electrons being shared. Oxygen has a double bond.)
7. What do the six scoops of ice cream represent in the illustration of N_2 on page 4? (Three bonded pairs of electrons represent a triple bond.)
8. Describe what you think is happening to the penguin in the CO_2 molecule in the illustration on page 4.
9. Name three things that the picture of CO_2 on page 4 illustrates about the molecule.
10. Describe what you think is happening to the penguins in the H_2O molecule in the illustration on page 4.
11. Explain what you think the crossed arrow represents in the comic book.
12. What are the two definitions of "dipole" given in the comic book?

Making Sense
What does electronegativity have to do with polarity?

If you finish early
Using polar bears and penguins, create an illustration showing a hydrogen sulfide molecule, H_2S. (Hint: You may wish to start with a Lewis dot structure.)

Making Sense Discussion   (15 min)

Major Goals: The processing of the comic book should focus on the main idea that electrons are not shared equally unless they are shared between identical atoms. **Electronegativity** should be defined and related to **polar covalent, nonpolar covalent,** and **ionic bonds.** Finally, examine how shape plays a role in determining the polarity of the entire molecule.

6. Define electronegativity.
When you get to the appropriate part of the discussion draw a picture of HCl on the board. Add the direction of the dipole and partial charges as you explain how the electrons are shared unequally.

Discussion goal:
Assist students in sharing their new understanding of polarity and how it relates to electronegativity.

Sample questions:
   List some of the things you learned about polarity from the comic book.
   How would you use polar bears and penguins to illustrate a polar molecule? A nonpolar molecule?
   According to the comic book, which elements tend to attract shared electrons to the greatest degree?

Points to cover:
When two atoms with different electronegativities are bonded, they tend to attract the bonded electrons to different degrees. This causes the electrons to spend more time around one of the atoms, resulting in a partial negative charge on this atom. This tendency of an atom to attract electrons shared between two atoms is called **electronegativity.** An atom that strongly attracts the shared electrons is considered **highly electronegative.** The atom with lower electronegativity will end up with a partial positive charge on it. The result is a polar bond. Chemists have a specific name for a molecule that has two poles — it is called a **dipole.** ("Di" means two.)

Using hydrogen chloride as an example, we can illustrate how the highly electronegative chlorine atom attracts the shared electrons to a greater degree, resulting in partial charges on the molecule.

251

TEACHER GUIDE

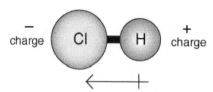

Hydrogen chloride - HCl
electrons pulled in this direction - direction of dipole

This illustration also uses a crossed arrow to show the **direction of the dipole** in HCl. The crossed end of the arrow indicates the positive (+) end of the polar bond and the arrow points in the direction of the negative (-) end.

Electronegativity values generally increase moving from left to right across the periodic table, thus the iceberg at the bottom of page 1 not only increases in thickness, but also has increasingly larger animals, including larger polar bears.

> Polar molecules are also called **dipoles.** The prefix di- means two. A dipole is a molecule with two partially charged ends, or poles. Chemists refer to polar molecules <u>as</u> dipoles and they also say that molecules with polar bonds <u>have</u> dipoles. This nomenclature can be a bit confusing with two related meanings for two closely-related meanings for the same word. wwwsameword.

7. Relate electronegativity to bonding.

Provide students with a general description of the three categories of bonds encountered in the comic book.

<u>Discussion goal:</u>
Assist students in differentiating between bonds that are nonpolar covalent, polar covalent, and ionic, depending on differences in electronegativity.

Sample questions:
How does the comic book illustrate the range of possible ways of sharing electrons?
What kind of charges form on the ions of elements that are the most electronegative? (they tend to form negative ions) What about the least electronegative elements? (they tend to form positive ions)
Why does the polar bear tell the penguin to try to be more positive? (the polar bear benefits more when the penguin is "more positive" – the more the polar bear attracts the shared pair of electrons, the more positive the partial charge on the penguin)

TEACHER GUIDE

Points to cover:

In a molecule such as H_2, the atoms are identical, so there is no difference in the degree to which each atom in the molecule attracts the shared electrons. These molecules are covalent and nonpolar. **Nonpolar covalent bonds** are the only bonds in which the electrons are truly shared equally. If the electronegativities between two atoms are even *slightly* different they form what is called a **polar covalent bond**. When the electronegativities between two atoms are greatly different the bond is called an **ionic bond**. In the case of an ionic bond the electron of one atom is *completely given up* to the other atom. The result of this kind of bond is two separate ions – one with a negative charge and the other with a positive charge. NaCl, sodium chloride (table salt), is an example of a compound with an ionic bond. Remember, just because an electron is completely transferred from one atom to another, does not mean that the bond between the two atoms is a weak one. Ionic bonds are quite strong.

8. Discuss how shape may relate to polarity.

Refer students to the illustrations of CO_2 on page 4 of the comic book. Explain how the shape of carbon dioxide results in a neutral, nonpolar molecule, even though the C=O bond is polar.

Discussion goal:

Assist students in interpreting the illustration of carbon dioxide in the comic book.

Sample questions:

What is happening to the penguin in the illustration of carbon dioxide on page 4 of the comic book?

What things does this illustration tell us about carbon dioxide? (There are double bonds between the C and each O; carbon is less electronegative than oxygen; the overall molecule is not polar.)

Although the electrons are not shared equally between C and O, CO_2 is a nonpolar molecule. Explain why.

Why is H_2O a polar molecule?

Points to cover:

The illustration of CO_2 shows a penguin being pulled between two large polar bears. The penguin represents a carbon atom in this drawing, and the polar bears are two oxygen atoms. The penguin is being pulled with equal strength or force in both directions. The fact that the penguin is being pulled means that the C=O bonds are polar. Nevertheless, they are pulling in equal and opposite directions. The net result is a neutral molecule. Thus, carbon dioxide is nonpolar even though it has two polar bonds. We will find more evidence later on that symmetrical molecules are often nonpolar due to their overall shape and conformation.

Students may notice that carbon and oxygen are next to each other on the periodic table and conceivably have electronegativities that are similar. According to the illustration at the bottom of page 1 of the comic book, both carbon and oxygen should be represented by polar bears. The carbon bear would be smaller and not as strong as the two oxygen bears. Apparently the artist used a penguin instead of a polar bear in order to make the point very clear that carbon has less attraction for bonded electrons than oxygen.

Check-in   (10 min)

9. Introduce the Check-in exercise.

Write the following exercise on the board for students to complete individually. Draw a picture of HI on the board as two circles linked together by a line. Label each atom (H and I). Students may wish to refer to the periodic table.

- Is the bond between these atoms polar? Explain your reasoning.
- How would the atoms be portrayed in the comic book – as polar bears, penguins, or both? Explain.

10. Discuss the Check-in exercise.

Get a sense of the level of understanding by collecting students' work or asking students to defend their choices.

Discussion goals:
Make sure that students have a grasp of polarity and electronegativity as they relate to the comic book.

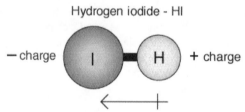

Sample questions:
Is this molecule polar? Explain. (Yes, the electrons are not shared evenly.)
Which atom(s) would be polar bears or penguins? (Hydrogen would be a penguin and iodine would be a polar bear.)
Which atom is more electronegative? Explain why you think so. (Iodine, because it is represented by a larger, stronger animal, the polar bear.)

On which end of a hydrogen iodide molecule would you find a partial negative charge? (iodine end)

In which direction are the electrons attracted? (towards iodine)

11. Wrap-up

Assist the students in summarizing what was learned in this class.

- Anytime there are two different types of atoms sharing electrons, there will be a partial negative charge on one atom and a partial positive charge on the other atom.
- Electronegativity measures the tendency of an atom to attract the electrons in a bond.
- The bonds are labeled nonpolar covalent, polar covalent, and ionic as the difference in electronegativity between the two atoms in the bond increases.

Homework

12. Assign homework.

Use the homework provided with the curriculum or assign your own.

Assign "The Bare Essentials of Polarity" (comic pages) for reading.

TEACHER GUIDE

Smells © UC Regents, LHS Living by Chemistry, 2003.   255

Homework – Investigation IV – Lesson 4

1. Read the comic "The Bare Essentials of Polarity." In your own words, explain the 3–5 main ideas. What do the polar bears represent? What do the penguins represent?

2. Using polar bears and penguins, create an illustration showing an ammonia molecule, NH_3. (Hint: You may wish to start with a Lewis dot structure.)

TEACHER GUIDE

ChemCatalyst

- Draw the Lewis dot structure for HCl.

- If the penguin represents a hydrogen atom and the polar bear represents a chlorine atom, what does the ice cream represent in the drawing? What do you think the picture is trying to illustrate?

- Do you think HCl will be attracted to a charged wand? Explain your thinking.

Smells © UC Regents, LHS Living by Chemistry, 2003.   257

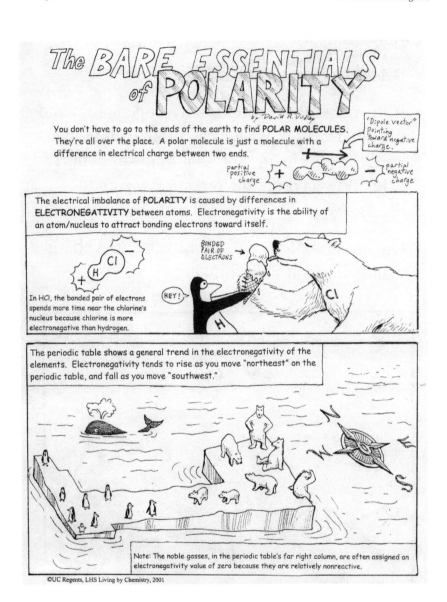

When two atoms with unequal electronegativity values bond, they do not share the bonding electrons evenly. The bonding electrons spend more time around the more electronegative atom, creating a PARTIAL NEGATIVE CHARGE on that atom. The other atom then has a PARTIAL POSITIVE CHARGE, and the bond is polar.

So the polarity of a bond is a function of the difference between the electronegativity values of two bonding atoms. Bonded atoms with equal electron-attracting strength will have nonpolar bonds.

However, if the electronegativity of two bonded atoms is unequal, then their bond will be polarized—maybe a little...

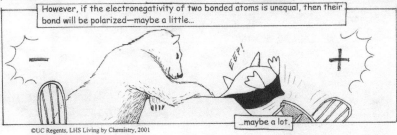

©UC Regents, LHS Living by Chemistry, 2001

259

TEACHER GUIDE

Because the elements have such varying electronegativities and can come together in so many different combinations, there is really a **CONTINUUM OF POLARITY IN BONDING**. For convenience, we can break the continuum down into three categories: (1) nonpolar covalent, (2) polar covalent, and (3) ionic.

NONPOLAR COVALENT

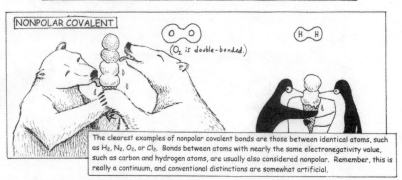

(O₂ is double-bonded.)

The clearest examples of nonpolar covalent bonds are those between identical atoms, such as H_2, N_2, O_2, or Cl_2. Bonds between atoms with nearly the same electronegativity value, such as carbon and hydrogen atoms, are usually also considered nonpolar. Remember, this is really a continuum, and conventional distinctions are somewhat artificial.

POLAR COVALENT

partial negative charge

partial positive charge

In a polar covalent bond, two atoms still share bonded pairs of electrons, but those electrons are decidedly more attracted to one atom than the other. Examples include bonds between carbon and oxygen atoms, or between hydrogen and fluorine atoms.

IONIC

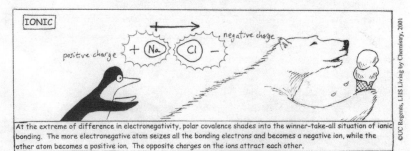

positive charge   negative charge

©UC Regents, LHS Living by Chemistry, 2001

At the extreme of difference in electronegativity, polar covalence shades into the winner-take-all situation of ionic bonding. The more electronegative atom seizes all the bonding electrons and becomes a negative ion, while the other atom becomes a positive ion. The opposite charges on the ions attract each other.

Polar bonds between atoms constitute **DIPOLES**. Actually, the word "dipole" can refer to several different things that are relevant here: (1) the polarity of an individual polar bond between atoms, (2) the net polarity of a polar molecule that may have several polar covalent bonds within it, and (3) the polar molecule itself.

Confusing? Let's look at some examples:

An N_2 molecule *isn't* a dipole (it's not a polar molecule), and it doesn't *have* any dipoles (polar bonds) within it.

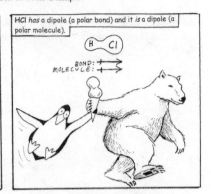

HCl *has* a dipole (a polar bond) and it *is* a dipole (a polar molecule).

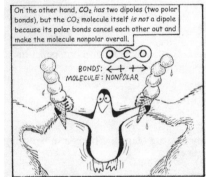

On the other hand, CO_2 *has* two dipoles (two polar bonds), but the CO_2 molecule itself *is not* a dipole because its polar bonds cancel each other out and make the molecule nonpolar overall.

Like CO_2, H_2O *has* two dipoles (two polar bonds). But because of H_2O's bent shape (caused by lone pairs of electrons on the oxygen atom), H_2O also has a dipole in the sense of an overall polarity. So H_2O *is* a dipole in the sense of being a polar molecule.

©UC Regents, LHS Living by Chemistry, 2001

The polarity of molecules can affect many of their other properties, such as their solubility, their boiling and melting points, and their odor.

Smells © UC Regents, LHS Living by Chemistry, 2003.   261

Polar Bears and Penguins

Name: _____

Period: _____ Date: _____

Purpose: In this lesson you will be exploring polarity and bonding between atoms in greater detail. A comic book will provide new information about these topics and will introduce you to the concept of electronegativity, which helps us to understand partial charges.

Use the comic book called "The Bare Essentials of Polarity" to answer the following questions.

1. How does the comic book define a "polar molecule?"

2. Define electronegativity as you understand it, after reading the first two pages of the comic book.

3. Interpret the picture at the bottom of page 1. Explain how the iceberg, penguins, and polar bears represent trends in electronegativity.

4. What is the artist trying to represent when there are two polar bears arm wrestling together, or two penguins arm wrestling together?

5. What three types of bonds are represented on page 3 of the comic book? What happens to the bonding electrons in each type of bond?

6. Explain why there are four scoops of ice cream in the illustration of O_2 on page 3.

7. What do the six scoops of ice cream represent in the illustration of N_2 on page 4?

8. Describe what you think is happening to the penguin in the CO_2 molecule in the picture on page 4.

9. Name three things that the picture of CO_2 on page 4 illustrates about the molecule.

10. Describe what you think is happening to the penguins in the illustration of H_2O on page 4.

11. Explain what you think the crossed arrow represents in the comic book.

12. What are the two definitions of "dipole" given in the comic book?

Making Sense

What does electronegativity have to do with polarity?

If you finish early. . .

Using polar bears and penguins, create an illustration showing a hydrogen sulfide molecule, H_2S. (Hint: You may wish to start with a Lewis dot structure.)

Smells © UC Regents, LHS Living by Chemistry, 2003.   263

Lesson Guide: Polar Bears and Penguins

Investigation IV – Lesson 4

In this lesson you will be exploring polarity and bonding between atoms in greater detail. A comic book will provide new information about these topics and will introduce you to the concept of electronegativity, which helps us to understand partial charges.

ChemCatalyst

Answer the following questions:

Consider the following illustration:

- Draw the Lewis dot structure for HCl.

- If the penguin represents a hydrogen atom and the polar bear represents a chlorine atom, what does the ice cream represent in the drawing? What do you think the picture is trying to illustrate?

- Would HCl be attracted to the charged wand? Explain your thinking.

Activity

Purpose: In this lesson you will be exploring polarity and bonding between atoms in greater detail. A comic book will provide new information about these topics and will introduce you to the concept of electronegativity, which helps us to understand partial charges.

Use the comic book called "The Bare Essentials of Polarity" to answer the following questions.

Smells © UC Regents, LHS Living by Chemistry, 2003.   85

1. How does the comic book define a "polar molecule?"

2. Define electronegativity as you understand it, after reading the first two pages of the comic book.

3. Interpret the picture at the bottom of page 1. Explain how the iceberg, penguins, and polar bears represent trends in electronegativity.

4. What is the artist trying to represent when there are two polar bears arm wrestling together, or two penguins arm wrestling together?

5. What three types of bonds are represented on page 3 of the comic book? What happens to the bonding electrons in each type of bond?

6. Explain why there are four scoops of ice cream in the illustration of O_2 on page 3.

7. What do the six scoops of ice cream represent in the illustration of N_2 on page 4?

8. Describe what you think is happening to the penguin in the CO_2 molecule in the picture on page 4.

9. Name three things that the picture of CO_2 on page 4 illustrates about the molecule.

10. Describe what you think is happening to the penguins in the illustration of H_2O on page 4.

11. Explain what you think the crossed arrow represents in the comic book. $\longleftarrow\!\!\!\!\!+$

12. What are the two definitions of "dipole" given in the comic book?

Making Sense Question:

What does electronegativity have to do with polarity?

If you finish early. . .

Using polar bears and penguins, create an illustration showing a hydrogen sulfide molecule, H_2S. (Hint: You may wish to start with a Lewis dot structure.)

Making Sense

When two atoms with different electronegativities are bonded together, they tend to attract the bonded electrons to different degrees. This causes the electrons to spend more time around one of the atoms, resulting in a partial negative charge on this atom. This tendency of an atom to attract electrons shared between two atoms is called **electronegativity**. An atom that strongly attracts the shared electrons is considered highly electronegative. The atom with lower electronegativity will end up with a partial positive charge on it. The result is a polar bond. Chemists have a specific name for a molecule that has two poles - it is called a **dipole**.

> Polar molecules are also called **dipoles**. The prefix di- means two. A dipole is a molecule with two partially charged ends, or poles. Chemists refer to polar molecules *as* dipoles and they also say that molecules with polar bonds *have* dipoles. These multiple definitions can be a bit confusing.

Check-in

Answer the following questions:

HI molecule

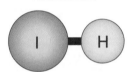

- Is the bond between these atoms polar? Explain your reasoning.

- How would the atoms be portrayed in the comic book – as polar bears, penguins, or both? Explain.

LIVING BY CHEMISTRY PROGRAM SAMPLER
©2006 Key Curriculum Press

STUDENT GUIDE

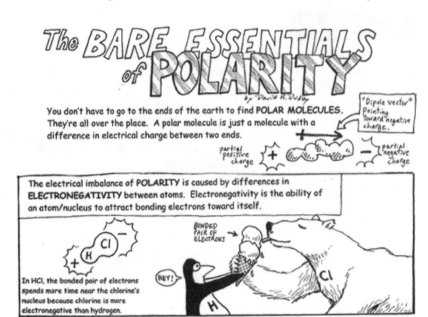

©UC Regents, LHS Living by Chemistry, 2001

STUDENT GUIDE

When two atoms with unequal electronegativity values bond, they do not share the bonding electrons evenly. The bonding electrons spend more time around the more electronegative atom, creating a PARTIAL NEGATIVE CHARGE on that atom. The other atom then has a PARTIAL POSITIVE CHARGE, and the bond is polar.

POLARITY IS UNFAIR!

OH, TRY TO BE MORE POSITIVE.

So the polarity of a bond is a function of the difference between the electronegativity values of two bonding atoms. Bonded atoms with equal electron-attracting strength will have nonpolar bonds.

However, if the electronegativity of two bonded atoms is unequal, then their bond will be polarized—maybe a little...

EEP!

...maybe a lot.

©UC Regents, LHS Living by Chemistry, 2001

Smells © UC Regents, LHS Living by Chemistry, 2003.   89

STUDENT GUIDE

Because the elements have such varying electronegativities and can come together in so many different combinations, there is really a **CONTINUUM OF POLARITY IN BONDING.** For convenience, we can break the continuum down into three categories: (1) nonpolar covalent, (2) polar covalent, and (3) ionic.

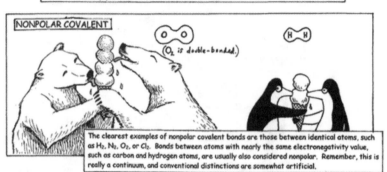

NONPOLAR COVALENT

(O₂ is double-bonded.)

The clearest examples of nonpolar covalent bonds are those between identical atoms, such as H₂, N₂, O₂, or Cl₂. Bonds between atoms with nearly the same electronegativity value, such as carbon and hydrogen atoms, are usually also considered nonpolar. Remember, this is really a continuum, and conventional distinctions are somewhat artificial.

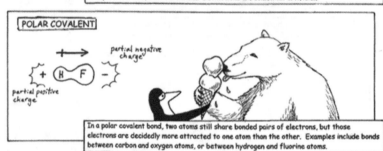

POLAR COVALENT

partial negative charge

partial positive charge

In a polar covalent bond, two atoms still share bonded pairs of electrons, but those electrons are decidedly more attracted to one atom than the other. Examples include bonds between carbon and oxygen atoms, or between hydrogen and fluorine atoms.

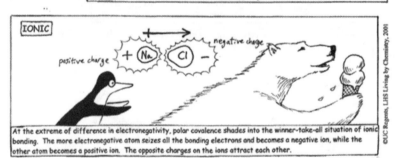

IONIC

positive charge

negative charge

At the extreme of difference in electronegativity, polar covalence shades into the winner-take-all situation of ionic bonding. The more electronegative atom seizes all the bonding electrons and becomes a negative ion, while the other atom becomes a positive ion. The opposite charges on the ions attract each other.

©UC Regents, LHS Living by Chemistry, 2001

LIVING BY CHEMISTRY PROGRAM SAMPLER
©2006 Key Curriculum Press

Polar bonds between atoms constitute **DIPOLES**. Actually, the word "dipole" can refer to several different things that are relevant here: (1) the polarity of an individual polar bond between atoms, (2) the net polarity of a polar molecule that may have several polar covalent bonds within it, and (3) the polar molecule itself.

Confusing? Let's look at some examples:

An N_2 molecule *isn't* a dipole (it's not a polar molecule), and it doesn't *have* any dipoles (polar bonds) within it.

HCl *has* a dipole (a polar bond) and it *is* a dipole (a polar molecule).

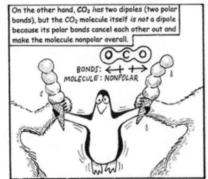

On the other hand, CO_2 *has* two dipoles (two polar bonds), but the CO_2 molecule itself *is not* a dipole because its polar bonds cancel each other out and make the molecule nonpolar overall.

Like CO_2, H_2O has two dipoles (two polar bonds). But because of H_2O's bent shape (caused by lone pairs of electrons on the oxygen atom), H_2O also has a dipole in the sense of an overall polarity. So H_2O *is* a dipole in the sense of being a polar molecule.

The polarity of molecules can affect many of their other properties, such as their solubility, their boiling and melting points, and their odor.

Smells © UC Regents, LHS Living by Chemistry, 2003.   91

Homework

Complete the following for homework:

1. Using polar bears and penguins, create an illustration showing an ammonia molecule, NH_3. (Hint: You may wish to start with a Lewis dot structure.)

STUDENT GUIDE

Materials for *Smells* Unit

KIT CONTENTS

100 g boiling stones (marble chips)

8 friction rods

100 plastic pipettes

120 4-dram vials with screw top

8 sets smells labels (13 per set)

300 cotton balls

24 string tags

6 12 × 12-in. wool pads

8 sets molecular models
(plus extras; 4200 pieces total)*

1 set space-filling molecular models
(6 per set)*; for building ball-and-stick
and space-filling models

Full set contains:

    420 glossy black spheres
    (4-hole carbon[B,S])

105 matte black spheres
(3-hole carbon[S])

900 white hemispheres (hydrogen[S])

50 red spheres (oxygen[B,S])

760 straight connectors
(single bonds[B])

20 curved connectors
(use in pairs for double bonds[B])

1395 short 1-disc connectors
(single bonds[S])

60 short 2-disc connectors
(double bonds[S])

8 sets Functional Group cards
(12 per set)

8 sets Space-Filling Models cards
(6 per set)

8 sets Molecules Card Sort cards
(22 per set); molecules A–V

2 sets Mystery Molecules cards
(22 per set); cards #1–22

For up-to-date information about kit contents, please visit www.keypress.com/chemistry .

---

*See pages xx–xxi for molecular model assembly instructions.

[B]for making ball-and-stick models

[S]for making space-filling models

MOLECULAR MODELING KITS

When it comes to understanding molecular structure, there is no substitute for hands-on experience with molecular models. Each *Smells* kit includes eight sets of modeling materials—enough for eight groups of students in a class. Each set contains enough parts to build several ball-and-stick or space-filling molecular models. Additional modeling sets may be purchased separately.

Ball-and-Stick Molecular Models

Preparation: Assemble these models ahead of time, or photocopy and hand out these instructions to teams to assemble their own set of models (recommended). You can label each model using masking tape or an index card attached to a string.

Assembly: To assemble the ball-and-stick models, use the structural formulas below. Black balls are used for carbon, white balls for hydrogen, and red balls for oxygen. The plastic sticks are used for single bonds. Two longer, curved plastic sticks are used for double bonds.

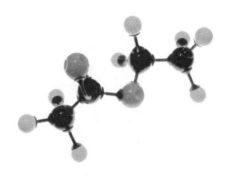

Molecule 1: Citronellol (C$_{10}$H$_{20}$O—sweet

- 10 carbon atoms, 20 hydrogen atoms, 1 oxygen atom
- 29 single bonds, 1 double bond

Molecule 2: Menthol (C$_{10}$H$_{20}$)—minty

- 10 carbon atoms, 20 hydrogen atoms, 1 oxygen atom
- 31 single bonds

Molecule 3: Fenchol (C$_{10}$H$_{18}$O)—camphor

- 10 carbon atoms, 18 hydrogen atoms, 1 oxygen atom
- 30 single bonds

Smells © UC Regents, LHS Living by Chemistry, 2003.   xx

...cular Models

...e the space-filling models, use the structural formulas and ... Black balls are used for carbon, the white balls for ...balls for oxygen. The short plastic connectors are used to ...single-disc connectors represent single bonds; the double-...ent double bonds. There are 2 types of black balls, with 3 ...The black balls with 4 holes are for carbon atoms with ...k balls with 3 holes are for the carbon atoms with one ...e-hole carbons have a matte finish so you can find them

Molecule 1: Citronellol (C$_{10}$H$_{20}$O)—sweet

- 8 4-hole and 2 3-hole carbon atoms, 20 hydrogen atoms, 1 oxygen atom
- 29 single bonds, 1 double bond

Molecule 2: Fenchol (C$_{10}$H$_{18}$O)—camphor

- 10 4-hole carbon atoms, 18 hydrogen atoms, 1 oxygen atom
- 30 single bonds

Molecule 3: Menthol (C$_{10}$H$_{20}$O)—minty

- 10 4-hole carbon atoms, 20 hydrogen atoms, 1 oxygen atom
- 31 single bonds

Smells © UC Regents, LHS Living by Chemistry, 2003.   xxi

OTHER MATERIALS NEEDED

Common lab ware, supplies, chemicals, and other easily obtainable materials have not been included in the Unit kit. They are listed below.

Your classroom should have a sink. Please note that Bunsen burners can be substituted for hot plates, but hot plates are safer to use.

Teachers should follow all state, local, and district guidelines concerning laboratory safety practices, safe handling of materials, and disposal of hazardous materials. In the absence of guidelines related to the handling, storage, and disposal of laboratory materials, teachers should refer to the Materials Safety Data Sheet (MSDS) for each chemical.

Essences

(These essences may be shared among classes. They will need to be replenished every 1 to 2 years.)

5 mL spearmint oil (L-carvone, minty)

5 mL fish oil (phenylethylamine, fishy)

5 mL pineapple flavor extract (isoamyl proprionate, sweet)

5 mL banana flavor extract (isoamyl acetate, sweet)

5 mL peppermint flavor extract (methone, minty)

5 mL apricot perfume oil (ethyl valerate, sweet)

5 mL butyric acid (butyric acid, putrid)

5 mL rum flavor extract (ethyl acetate, sweet)

5 mL rose perfume essence (citronellol, sweet)

5 mL Canadian fir balsam oil (fenchol, camphor)

5 mL jasmine perfume oil (geraniol, sweet)

5 mL mint flavor extract (menthol, minty)

5 mL pine oil (borneol, camphor)

Other Supplies

(These quantities are for a class of 32 students.)

mint leaves

fruit Chapstick® (1 stick)

apple tea (1 teabag)

Altoids® (8 tablets)

eucalyptus or bay leaves

rose water (5 mL)

Wint-O-Green LifeSavers® (1 per student)

gumdrops (4–5 per pair of students)

miniature marshmallows (12 per pair of students)

toothpicks (15 per pair of students)

organic molecule models or toothpicks and gumdrops/marshmallows (optional) scissors (one pair per student)

32 3 × 3-in. waxed paper squares

7 soft cloths (flannel or silk)

8 glue sticks

distilled water

Other Chemicals

(These quantities are for a class of 32 students.)

10 mL isoamyl alcohol

10 mL ethyl alcohol

10 mL n-butyl alcohol

10 mL acetic acid

1 mL concentrated 18 M sulfuric acid

20 g maleic acid

20 g fumaric acid

50 mL hexane

5 mL thymol blue indicator

50 mL vinegar

50 mL rubbing alcohol

8 2-cm magnesium strips

8 crystals of sodium carbonate (washing soda)

Lab Ware

(These quantities are for a class of 32 students.)

32 pairs of safety goggles

8 hot plates

8 large well plates (at least 6 wells)*

8 spatulas

32 droppers

7 burets and 4–7 buret stands (1 for demonstration; 2 per station)

24 0.5-mL microscale test tubes*

16 5-mL test tubes*

8 50-mL beakers

7 500-mL beakers

safety gear for mini-drama (optional)

*Note: Well plates are included in the *Toxins* Lab Kit and can substitute for microscale test tubes as well.

LESSON-BY-LESSON MATERIALS GUIDE

An attempt has been made to provide the specialty items needed for each lab. However, common lab ware, chemicals, and other easily obtainable materials have not been included in this kit. They are listed in the Other Materials Needed column.

For consumable items, one kit contains enough materials for ten classes of 32 students. Please note that Bunsen burners can be substituted for hot plates, but hot plates are safer to use.

Lesson I-1: Cat Food and Stinky Cheese

Preparation: Essences can be obtained from a grocery store, a bath shop, or a chemical supplier.

You may choose to set up stations for this activity, in which case you will need to set up only one or two vials with each smell.

Materials Provided in Kit	Other Materials Needed
40 smell vials	spearmint oil (L-carvone, minty)
40 cotton balls	fish oil (phenylethylamine, fishy)
	pineapple flavor extract (amyl proprionate, sweet)
	banana flavor extract (isoamyl acetate, sweet)
	peppermint flavor extract (menthone, minty)
	mint leaves
	fruit Chapstick®
	apple tea
	Altoids®
	eucalyptus or bay leaves
	rose water

Lesson I-2: Sniffing Around

Materials Provided in Kit	Other Materials Needed
24 smell vials	apricot perfume oil (ethyl valerate, sweet)
24 cotton balls	butyric acid (butyric acid, putrid)
	rum flavor extract (ethyl acetate, sweet)

Lesson II-1: Molecules in Two Dimensions

Materials Provided in Kit	Other Materials Needed
8 sets Functional Group cards A–L (12 per set)	none

Lesson II-7: Create a Smell

Preparation: Be extremely careful when handling sulfuric acid and have baking soda available to neutralize spills. Use hot plates instead of Bunsen burners; organic chemicals are flammable! Place chemicals in reagent bottles at each station and close all bottles or containers when not in use because of fire danger with alcohol.

Materials Provided in Kit	Other Materials Needed
boiling stones (a few grams per group)	5 mL butyric acid (5 drops per group)
24 disposable pipettes	10 mL isoamyl alcohol (10 drops per group)
	10 mL ethyl alcohol (10 drops per group)
	10 mL n-butyl alcohol (10 drops per group)
	10 mL acetic acid (10 drops per group)
	8 50-mL beakers
	8 hot plates
	24 0.5-mL microscale test tubes
	1 mL of concentrated 18 M sulfuric acid (1 drop per group)
	32 pairs of safety goggles

Disposal: Have waste containers available for the esters produced in this lab. Make sure hot plates are turned off when lab is finished.

Lesson III-1: New Smells, New Ideas

Materials Provided in Kit	Other Materials Needed
40 smell vials	rose perfume essence (citronellol, sweet)
40 cotton balls	Canadian fir balsam oil (fenchol, camphor)
	jasmine perfume oil (geraniol, sweet)
	mint flavor extract (menthol, minty)
	pine oil (borneol, camphor)

Lesson III-2: Molecules in Three Dimensions

Materials Provided in Kit	Other Materials Needed
ball-and-stick molecular models (1 set per group)	none
Each group will need one set containing:	
50 glossy black spheres (4-hole carbon)	
58 white spheres (hydrogen)	
6 red spheres (oxygen)	
90 straight connectors (single bonds)	
2 curved connectors (use in pairs for double bonds)	

Lesson III-3: Two's Company

Preparation: Prepare one set of simple molecules using the molecular model pieces—methane (CH_4), ammonia (NH_3), and water (H_2O). (Use any 4-hole ball for nitrogen, and tell students it represents a nitrogen atom here.)

Materials Provided in Kit	Other Materials Needed
ball-and-stick models of methane, ammonia, and water (1 set of 3)	gumdrops (4–5 per pair of students)
	miniature marshmallows (12 per pair of students)
	toothpicks (15 per pair of students)

Lesson III-4: Let's Build It

Preparation: Assemble one set of molecules.

Materials Provided in Kit	Other Materials Needed
ball-and-stick models of methane, ammonia, water, hydrogen fluoride, and neon with lone pairs represented (1 set of 5)	organic molecule models or toothpicks and gumdrops/marshmallows (optional)
8 ball-and-stick model kits (1 kit per group)	

Lesson III-5: Shape Matters

Preparation: Assemble one set of molecules (see lesson for molecular structure). Cut magnesium into 2-cm strips.

Materials Provided in Kit	Other Materials Needed
ball-and-stick models of maleic acid and fumaric acid (1 set of 2)	20 g maleic acid
	20 g fumaric acid
	5 mL thymol blue indicator
	8 2-cm magnesium strips
	8 crystals of sodium carbonate (washing soda)
	16 5-mL test tubes
	8 spatulas
	distilled water
	8 large well plates (with at least 6 wells)

Lesson III-6: What Shape Is That Smell?

Materials Provided in Kit	Other Materials Needed
space-filling molecular models	safety gear for mini-drama (optional)
Each group will need one set containing:	
48 glossy black spheres (4-hole carbon)	
12 matte black spheres (3-hole carbon)	
106 white hemispheres (hydrogen)	
165 short 1-disc connectors (single bonds)	
7 short 2-disc connectors (double bonds)	
8 sets Space-Filling Models cards (6 per set)	
ball-and-stick model of citronellol	

Lesson III-7: Sorting It Out

Materials Provided in Kit	Other Materials Needed
8 sets Molecules Card Sort cards (22 per set)	none
2 sets Mystery Molecules cards (22 per set)	

Lesson IV-1: Breaking Up Is Hard to Do??

Materials Provided in Kit	Other Materials Needed
none	8 pairs of scissors
	8 glue sticks

Lesson IV-3: Attractive Molecules

Preparation: Obtain soft cloths ahead of time.

Materials Provided in Kit	Other Materials Needed
7 plastic or Lucite wands (or plastic rulers)	50 mL hexane
6 12 × 12-in. wool pads	7 burets and 4–7 buret stands (1 for demonstration, 2 per station)
	7 500-mL beakers
	32 3 × 3-in. waxed paper squares
	50 mL water
	50 mL vinegar
	50 mL rubbing alcohol
	7 soft cloths (flannel or silk)
	32 droppers

Disposal: All materials can be saved and reused.

Lesson IV-6: I Can Relate

Materials Provided in Kit	Other Materials Needed
none	scissors (1 pair per student)

Lesson IV-8: Take a Deep Breath

Materials Provided in Kit	Other Materials Needed
none	Wint-O-Green LifeSavers® (1 per student)

WEATHER

Gas Laws and Phase Changes

Contents

Main Topics Covered

COVERED IN DEPTH

Mole/number conversions

Gas pressure

Random motion of gas particles

Gas laws

STP

Temperature scales

Absolute zero

Kinetic theory of gases

Ideal gas law

Temperature and heat

PREREQUISITES FOR THE UNIT

Periodic table

Atomic number

Atomic mass

Investigation Summaries

Weather emphasizes quantitative analysis and provides ample opportunities for using mathematical functions to model scientific phenomena. This unit uses weather patterns to investigate physical change and topics like air pressure and the gas laws.

INVESTIGATION I: LOCATING MATTER

The class now turns its attention from molecular structure to physical change, using weather patterns and the water cycle to investigate volume, mass, and density. In Investigation I students compare methods of rain collection by graphing volume and height, and they calculate the snowpack for a given year. They also review some of the math that will be necessary in future lessons.

INVESTIGATION II: HEATING MATTER

The *Weather* unit continues by studying heat and its effect on matter in different physical states. Beginning with students building their own thermometers, this investigation studies the melting and boiling points of water, Charles' law, and specific heat capacity.

INVESTIGATION III: MOVING MATTER

This investigation considers air pressure and its relationship to volume and temperature. Students use air in a syringe to learn about Boyle's and Gay-Lussac's laws, and eventually connect their knowledge to wind and weather patterns.

INVESTIGATION IV: COUNTING MATTER

Investigation IV looks at gases on a molecular level, investigating mass and number of particles. Students study air pressure and learn how it connects to density, and they study Avogadro's number and the ideal gas law. They also begin to explore humidity and its effects on the weather.

BEFORE CLASS...

LESSON 2 – Feeling Under Pressure

Key Ideas:

Pressure is measured as the force per unit of area. The pressure of a gas is inversely proportional to the volume the gas occupies. In other words, when one variable gets larger, the other variable gets smaller, and vice versa.

What Takes Place:

This is a two-part activity. Students first explore pressure by covering the end of a syringe with their fingertip and pushing the plunger. They work in pairs for this portion of the activity and answer a few worksheet questions. Part II of the activity involves using a syringe that has been capped off with air trapped inside the chamber. Several student volunteers work to push the plunger down on a syringe that is placed on top of a bathroom scale. The weight is read off the scale and reflects the force required to push the syringe to a certain volume. Both the volume and the weight data are collected by the class and graphed.

Materials:

- Student worksheet
- Two medium-sized party balloons
- Approximately two cups of sand
- 16 50-mL plastic syringes (without caps on the tips)
- 1 50-mL plastic syringes per bathroom scale (with caps securely glued onto the tips)
- bathroom scale – one mandatory, more if available

TEACHER GUIDE

Investigation III – Moving Matter
LESSON 2 – Feeling Under Pressure

The purpose of today's class is to explore the relationship between pressure and volume of gases. Gas pressure and volume exist in inverse relationship to one another (given constant temperature). Students are first introduced to the definition of pressure as force per unit of area. Students then experiment with squeezing air in a closed container (a syringe) measuring the force required by using a bathroom scale. They graph and begin to interpret the results.

Exploring the Topic   (5–10 min)

1. Introduce the ChemCatalyst exercise.

Prepare two medium-sized party balloons before class. Blow one up to a fairly large volume with air. Use a funnel to fill the other one with as much sand as you can get into it. (It will still be smaller than the first balloon.) Show students the balloons and write the following questions on the board for them to complete individually.

- Which balloon contains the greatest volume of material? Explain why you think so.
- Which balloon weighs more?
- Which material is exerting more pressure on the walls of the balloon? How can you tell?

2. Discuss the ChemCatalyst exercise.

Use the discussion to get a sense of students' initial ideas.

Discussion goals:
Assist the students in sharing their initial ideas about air pressure and the pressure due to the weight of an object.

Sample questions:
Which balloon has the greater volume of material?
Which balloon weighs more?
Why is the sand balloon flat on the bottom?
How is pressure measured?

The balloon full of air occupies a greater volume. The sand weighs more. However, the pressure from the air molecules on the inside walls of the balloon is quite high and causes the balloon to stretch considerably. The sand balloon is flat on the bottom because the weight of the sand is pushing down on the balloon. The air balloon has equal pressure on all parts of the walls, thus the balloon is rounded.

TEACHER GUIDE

3. Introduce measurement of pressure.

Points to cover:

If you were to hold each balloon in your hand you could feel the different pressures being exerted by these two different substances. The force from the sand would push straight down against your hand. The force from the air can be felt in the extreme tightness of the balloon.

Pressure is defined as a force per unit area. The downward pressure exerted by the sand balloon is greater than the downward pressure exerted by the air balloon. This is because the sand balloon has a greater mass and weighs more. However, gas pressure is due to the collisions of molecules against a surface. The air inside the air balloon exerts a greater pressure on the inside surface of the balloon compared with the sand. This is because the air molecules are moving fast and colliding with the inside walls of the container. Thus, the pressure inside the air balloon is greater than the pressure inside the sand balloon and the balloon is stretched tight.

Pressure is measured differently in these two cases. The sand balloon exerts a downward pressure. The force in this case is the weight of the balloon and its contents due to gravity (F = mass x gravity). The weight of the sand is distributed over the bottom of the balloon.

The air molecules exert a pressure on the walls of the air balloon. The force is due to the rapidly moving air molecules colliding with the walls of the balloon (F = mass x acceleration). The forces due to the moving air molecules are distributed over the inside surface of the balloon.

Because pressure is defined as force per unit of area, air pressure can also be measured in pounds/inch^2. The air pressure in tires is measured in psi, or pounds per square inch. Remember, the pressure exerted by the atmosphere all around us is measured in atm, or atmospheres. We will learn more about measuring pressure at a later time.

> **Pressure** is defined as the force per unit area.

4. Explain the purpose of the activity.

If you wish you can write the main question on the board.

Points to cover:

We have already discovered that a mathematical relationship exists between the temperature and volume of a gas (Charles's Law). In today's class we will be using downward pressure (weight) per area to measure the pressure of some trapped air. Tell students they will be answering the question: "What is

Weather © UC Regents, LHS Living by Chemistry, 2003.

TEACHER GUIDE

the relationship between pressure and volume for gases (if we keep the temperature and amount of gas constant)?"

Activity – Feeling Under Pressure   (15 min)

5. Explain the procedure for Part I. (Worksheet)

Tell students that they will be using a syringe as a container with a volume that can change. To begin, they will make pressure observations while using their fingertip to cover the bottom of the syringe. After passing out materials allow students about 5 to 10 minutes to experiment with the syringe and answer the questions in Part I of the worksheet before proceeding to Part II as an entire class.

Materials (for each pair of students)
16 50-mL plastic syringes (uncapped)

Part I: Qualitative observations
1. Work in pairs.
2. Cover the tip of the syringe with your fingertip. Be sure to make a good seal.
3. Record the volume of gas in the syringe, using the scale on the syringe.
4. Apply pressure to the syringe, causing the plunger to read 40 mL.
5. Apply more pressure to the syringe, so that the inside volume is 30 mL, 20 mL, and so on.
6. Record your observations and answer the questions below.

Answer the following questions:
1. What did you feel when you pushed the plunger down from 40 mL to 30 mL to 20 mL. (It is harder to go from 30 mL to 20 mL compared with 40 mL to 30 mL.)

2. Are you able to push the plunger all the way to the bottom? Explain why or why not. (No, because there is still air inside the syringe and it pushes back.)

3. Does the amount of air inside the syringe change? Explain your thinking. (No, because the air cannot escape from the inside with your finger sealing the tip.)

6. Collect weight vs. volume data.

Ask a couple of volunteers to perform similar explorations with the syringe, except that the tip of the syringe will be placed on a bathroom scale. For this part of the experiment, the cap provided with the syringe needs to be sealed to the tip of the syringe with epoxy so that air cannot escape.

Weather © UC Regents, LHS Living by Chemistry, 2003.   169

Materials:

1 50-mL plastic syringe (or more if available, with the caps sealed to the tip with epoxy)

1 bathroom scale (or more if available)

Caution: The cap on the tip of the syringe should always be pointed down, away from eyes.

Part II: Do the following as a demonstration:

1. Work as a class, with several volunteers performing the experiment. (Three volunteers works well for this demonstration. If you have more materials you may have more than one group completing these measurements at once.)

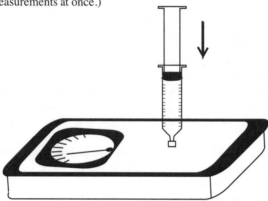

2. Hold the syringe with the tip on top of a bathroom scale.
3. One student should depress the plunger slightly.
4. A second student should read the volume.
5. A third student should read the number of pounds that is exerted on the bathroom scale.
6. All students should record the data in the table provided.
7. Repeat for at least 5 volumes.

Sample Data:

Trial	Volume (mL)	Weight (lbs)	Pressure (lbs per in^2)
1	50 mL	0	0
2	40 mL	10	3.2
3	30 mL	15	4.8
4	20 mL	25	8.0
5	15 mL	35	11.1
6	10 mL	55	17.5

Weather © UC Regents, LHS Living by Chemistry, 2003.

Answer the following questions:

1. Fill in the table with the results of the demonstration.

2. Convert the weight to pressure by dividing by the area to get pounds per square inch. The area is π times the radius of the syringe squared (Area = πr^2) (For a 2 cm diameter syringe, the area = $\pi(1)^2 = 3.14$ cm^2)

3. Make a graph of pressure versus volume. Put pressure on the y axis and volume on the x-axis.

4. Describe what happens to the pressure as the volume decreases. (As the volume decreases, the pressure increases)

5. Use you graph to estimate the following:

   a) If the volume is reduced to 32 mL what will the pressure be?

   b) If the volume is reduced to 16 mL what will the pressure be?

6. Explain why the number of molecules in the syringe didn't change, but the volume did. (No molecules could escape. The volume changed because of pressure applied to the plunger of the syringe.)

7. Describe the relationship between pressure and volume. (Pressure is inversely proportional to volume. When pressure increases, volume decreases.)

Making sense:
Why is it so difficult to push the plunger in as the volume gets smaller?

Making Sense Discussion   (15 min)

Major goals: This lesson will be debriefed in more detail in the following lesson. For now it is important that students get an opportunity to share their experiences with the syringe. They take a look at the graph and interpret its outcome, noting the inverse relationship between pressure and volume. If you wish you can have students correct for the additional pressure on the syringe due to the atmosphere, however this is not mandatory.

7. Briefly discuss Part I of the activity.

Discussion goals:
Discuss what happens to the air inside the syringe as the plunger is pushed in.

Sample questions:
   What happens to the air on the inside of the syringe as the plunger is pushed in? (Does the amount of air change?)

TEACHER GUIDE

Explain why the volume of air can change even if the number of air molecules does not change.

Can you decrease the volume to zero? Why or why not?

Why does it get harder and harder to depress the plunger as the volume gets smaller and smaller?

Students should observe that it becomes harder and harder to depress the plunger as the volume gets smaller. The air inside the syringe is being compressed into a smaller and smaller space. Since no air is escaping, the number of air molecules does not change, only the volume the molecules occupy. In other words, the molecules in the air are getting closer to one another. All matter occupies some space, so the volume inside cannot go to zero. Indeed, you cannot even get close to zero because the air molecules are "pushing back" with high pressure on the walls of the container due to their motions.

8. Briefly discuss the weight changes.

Draw a graph on the board and ask student volunteers to graph the data that was collected today.

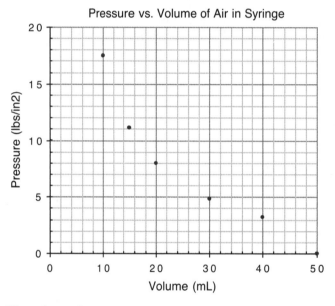

Pressure vs. Volume of Air in Syringe

<u>Discussion goals:</u>

Discuss how the weight on the scale changes as the volume of air in the syringe is decreased.

Sample questions:

What does the graph of your data look like? (a curve, does not begin at the origin)

Why do you think the graph is not a straight line? (it takes more and more force the smaller the volume)

Weather © UC Regents, LHS Living by Chemistry, 2003.

TEACHER GUIDE

Explain what happens to the weight on the scale as you depress the plunger.
   (the weight increases)
How did you calculate the pressure from the weight?
Use your graph to estimate the pressure for 32 mL and 16 mL.
Describe how the pressure varies as the volume gets smaller. (the pressure
   increases more dramatically as the volume gets smaller)
Why is it so difficult to push the plunger in as the volume gets smaller?
   (because the air molecules are colliding more frequently with the walls of
   the container, and hence the pressure is increasing)

Note: The points on the graph are not exactly those expected for the equation P = constant
(1/V). This is because the pressures need to be corrected for the fact that when the weight on
the bathroom scale measures zero, there is 1 atm pushing down on the balance. In other
words, when the balance reads zero, it is still "weighing" the atmosphere. The correction for
atmospheric pressure is given below should you choose to use it. For now, focus on the
inverse relationship between pressure and volume.

Points to cover:
The graph reflects the fact that as the volume of the syringe decreases it takes
more and more force to push the molecules of air into that space. The graph
also reflects the fact that the relationship between pressure and volume is an
inverse one. In other words, as the value of one gets bigger the value of the
other gets smaller.

> Gas pressure and volume have an **inverse** relationship. When the volume of a
> given amount of gas is decreased, its pressure increases. When the volume of a
> given amount of gas is increased, its pressure decreases.

Inform students that you will examine the data from this activity in more
detail in the next lesson.

Students should observe that as the volume is decreased, the weight increases. However, it is
easier to push the plunger in a little bit. The more the plunger is pushed in, the smaller the
volume is for the same amount of gas to occupy and the harder it is to push in.

9. Briefly discuss the correction due to atmospheric pressure.

This portion of the discussion is optional.

Points to cover:
When the scale reads zero pounds there is one atmosphere pressure on the
scale and on the syringe. This is because the atmosphere around us always
exerts a force of one atmosphere (at sea level). So, any measurement taken
using the bathroom scale will have to have one atmosphere added at the end.
By the way, 1 atm = 14.7 lbs/in^2.

The weight is the force applied over the area of the plunger. Thus, it is
necessary to divide the weight that was measured by the cross-sectional area

of the plunger in order to get lbs/in^2. You can accomplish this by measuring the diameter of the syringe in inches and using Area $= \pi \, r^2$.

$$\text{Pressure} = \frac{\text{weight in pounds}}{\text{area of cross section of syringe}} + 14.7 \text{ lbs/in}^2$$

Check-in   (5 min)

10. Complete the Check-in exercise.

Write the following question on the board for students to complete individually.

Imagine you have a plastic bottle that is capped. It contains nothing but air.

- What happens to the volume the bottle if you squeeze the bottle tightly?
- What happens to the pressure inside the bottle when it is squeezed?

11. Discuss the Check-in exercise.

Get a sense of the level of understanding by asking students to defend their choices.

Discussion goals:
Check to see that students understand the relationship between pressure and volume of air inside a container.

12. Wrap-up

Assist the students in summarizing what was learned in this class.
- Pressure and volume have an inverse relationship, that is, when one gets larger, the other gets smaller.

Homework

13. Assign the following for homework.

Use the homework provided with the curriculum or assign your own.

TEACHER GUIDE

Homework – Investigation III – Lesson 2

1. You blow up a balloon before going SCUBA diving. You put on your gear and descend to 30 ft with the balloon. The total pressure from the ocean at that depth is measured at 2 atmospheres (2 atm). Assume temperature is constant.

   a)   What happens to the volume of the balloon?

   b)   What happens to the pressure of the air on the inside of the balloon?

2. You are filling a bicycle tire with air using a bicycle pump. It gets harder and harder to push the plunger on the pump the more air is in the tire. Explain what is going on.

TEACHER GUIDE

Feeling Under Pressure

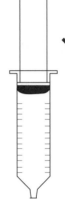

Name: _____

Period: _____ Date: _____

Purpose: This activity allows you to compare the pressure and volume of a sample of air.

Part I: Experimenting with pressure.
1. Work in pairs.
2. Cover the tip of the syringe with your fingertip. Be sure to make a good seal.
3. Record the volume of the syringe, using the scale on the side of the syringe.
4. Apply pressure to the syringe, causing the plunger to read 40 mL.
5. Apply more pressure to the syringe so that the inside volume is 30 mL, 20 mL, and so on.
6. Record your observations and answer the questions below.

Answer the following questions:
1. What did you feel when you pushed the plunger down from 40 mL to 30 mL to 20 mL?

2. Are you able to push the plunger all the way to the bottom? Explain why or why not.

3. Does the amount of air inside the syringe change? Explain your thinking.

Part II : Quantitative data.
You will collect quantitative data as a class.

1. Fill in the table with the results of the demonstration.

2. Convert the weight to pressure by dividing by the area to get pounds per square inch. The area is the radius of the syringe squared times π (Area = πr^2)

Trial	Volume (mL)	Weight (lbs)	Pressure (lbs per in^2)
1			
2			
3			
4			
5			
6			

Weather © UC Regents, LHS Living by Chemistry, 2003.

3. Make a graph of pressure versus volume. Put pressure on the y axis and volume on the x axis.

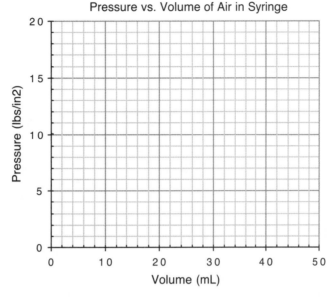

4. Describe what happens to the pressure shown on the scale as the volume of the air decreases.

5. Use your graph to estimate the following:

   a) If the volume is reduced to 32 mL what will the pressure be?

   b) If the volume is reduced to 16 mL what will the pressure be?

6. Explain why the number of molecules in the syringe didn't change, but the volume did.

7. Describe the relationship between pressure and volume.

Making sense:
Why is it so difficult to push the plunger in as the volume gets smaller?

Weather © UC Regents, LHS Living by Chemistry, 2003.   177

TEACHER GUIDE

Lesson Guide: Feeling Under Pressure

Investigation III – Lesson 2

In today's class you will explore the relationship between pressure and volume of gases.

ChemCatalyst

Your instructor will show you two balloons containing different substances. Answer the following questions:

- Which balloon contains the greatest volume of material? Explain why you think so.
- Which balloon weighs more?
- Which material is exerting more pressure on the walls of the balloon? How can you tell?

After discussing the ChemCatalyst exercise the instructor will introduce you to the measurement of pressure.

Activity

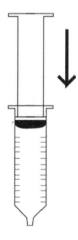

Materials: (for each pair of students)
16 50-mL plastic syringes (uncapped)

Part I: Qualitative observations

1. Work in pairs.

2. Cover the tip of the syringe with your fingertip. Be sure to make a good seal.

3. Record the volume of gas in the syringe, using the scale on the syringe.

4. Apply pressure to the syringe, causing the plunger to read 40 mL.

5. Apply more pressure to the syringe, so that the inside volume is 30 mL, 20 mL, and so on.

6. Record your observations and answer the questions below.

STUDENT GUIDE

Answer the following questions:

1. What did you feel when you pushed the plunger down from 40 mL to 30 mL to 20 mL.

2. Are you able to push the plunger all the way to the bottom? Explain why or why not.

3. Does the amount of air inside the syringe change? Explain your thinking.

Part II: Collecting weight vs. volume data.

This portion of the activity will be completed as an entire class. Student volunteers will complete the experiment. Make sure to record the data that is collected.

Materials:
1 50-mL plastic syringe (or more if available, with the caps sealed to the tip with epoxy)
1 bathroom scale (or more if available)

Caution: The cap on the tip of the syringe should always be pointed down, away from eyes.

1. Work as a class, with several volunteers performing the experiment.

2. Hold the syringe with the tip on top of a bathroom scale.

3. One student should depress the plunger slightly.

4. A second student should read the volume.

5. A third student should read the number of pounds that is exerted on the bathroom scale.

6. All students should record the data.

7. Repeat for at least 5 volumes.

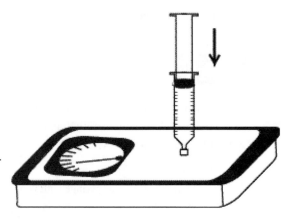

Complete the following for Part II:

1. Create a table similar to the one below. Fill in the table with the results of the demonstration.

2. Convert the weight to pressure by dividing by the area to get pounds per square inch. The area is the radius of the syringe squared times □ (Area = □r²).

Weather© UC Regents, LHS Living by Chemistry, 2003

LIVING BY CHEMISTRY PROGRAM SAMPLER
©2006 Key Curriculum Press

Trial	Volume (mL)	Weight (lbs)	Pressure (lbs per in^2)
1			
2			
3			
4			
5			
6			

Copy this table into your notebook

3. Create a graph similar to the one below of pressure versus volume. Put pressure on the y axis and volume on the x axis.

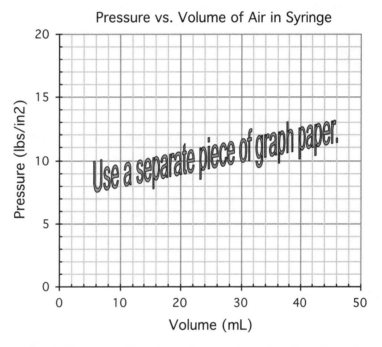

Pressure vs. Volume of Air in Syringe

Use a separate piece of graph paper.

4. Describe what happens to the pressure shown on the scale as the volume of the air decreases.

5. Use your graph to estimate the following:

   a) If the volume is reduced to 32 mL what will the pressure be?

   b) If the volume is reduced to 16 mL what will the pressure be?

6. Explain why the number of molecules in the syringe didn't change, but the volume did.

STUDENT GUIDE

7. Describe the relationship between pressure and volume.

Making Sense Question:
Why is it so difficult to push the plunger of the syringe in as the volume gets smaller?

Check-in

Answer the following questions:

Imagine you have a plastic bottle that is capped. It contains nothing but air.

*   What happens to the volume of the bottle if you squeeze the bottle tightly?

*   What happens to the pressure inside the bottle when it is squeezed?

Homework

Complete the following for homework:

1. You blow up a balloon before going SCUBA diving. You put on your gear and descend to 30 ft with the balloon. The total pressure from the ocean at that depth is measured at 2 atmospheres (2 atm). Assume temperature is constant.

    a) What happens to the volume of the balloon?

    b) What happens to the pressure of the air on the inside of the balloon?

2. You are filling a bicycle tire with air using a bicycle pump. It gets harder and harder to push the plunger on the pump, the more air there is in the tire. Explain what is going on.

STUDENT GUIDE

Materials for *Weather* Unit

KIT CONTENTS

Lab Ware

8 squirt bottles

16 1.5-mm capillary tubes closed at one end

16 small glass vials

16 rubber stoppers with hole

16 50-mL syringes without caps

2 50-mL syringes with screw caps

3 ft of clear tubing

24 ft of flexible tubing

10 plastic pipettes

Other Supplies

16 permanent markers

8 wet-erase markers

16 12-in. transparent rulers, marked in inches and cm

32 small rubber bands

Laminated index card

330 party balloons[R]

320 short straws to fit snugly in tubing

Food coloring[R]

Long matches[R]

2 clear plastic cups

400 mL rubbing alcohol

16 oven mitts

8 vials mineral oil

1 cup potting soil

2 cups sand

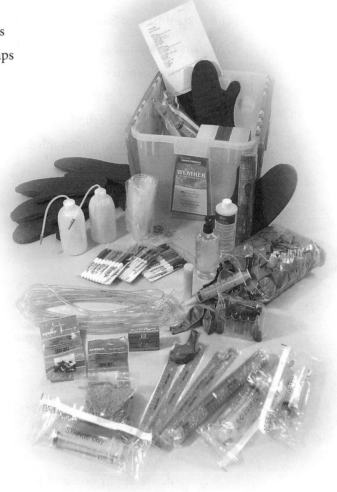

For up-to-date information about kit contents, please visit   www.keypress.com/chemistry   .

[R]Consumable; will need to be replenished. This kit includes enough materials for approximately 10 classes of 32 students, or 2–3 years.

OTHER MATERIALS NEEDED

Common lab ware, supplies, chemicals, and other easily obtainable materials have not been included in the Unit kit. They are listed below.

Your classroom should have a sink. Please note that Bunsen burners can be substituted for hot plates, but hot plates are safer to use.

Teachers should follow all state, local, and district guidelines concerning laboratory safety practices, safe handling of materials, and disposal of hazardous materials. In the absence of guidelines related to the handling, storage, and disposal of laboratory materials, teachers should refer to the Materials Safety Data Sheet (MSDS) for each chemical.

Lab Ware

(These quantities are for a class of 32 students.)

32 pairs of safety goggles (1 per student)

16 250-mL Erlenmeyer flasks

1000-mL Erlenmeyer flask

16 25-mL graduated cylinders

8 graduated cylinders (preferably 250-mL or 500-mL)

16 buckets or large beakers

16 100-mL beakers

2 400-mL beakers

24 250-mL beakers

16 test tubes

16 test tube holders

16 thermometers (Celsius scale)

1 pair of tongs (4–6 more tongs optional)

Ring stand and clamp

Large, shallow tub to catch spills

Flasks, stoppers, glass tubing, and rubber tubing (if vacuum pump is unavailable)

8 scale balances

8 hot plates

Vacuum desiccator or side-arm flask with stopper

Funnel

Wooden or stiff plastic ruler

Other Equipment

Vacuum pump (recommended)

Bathroom scale (more scales optional)

1 Styrofoam cooler or ice chest

1 hammer

2–4 5-gallon cylindrical containers (such as plastic flower pots)

8–16 buckets or large beakers

Desk lamp with high-intensity bulb

Household Items

17 2-liter plastic bottles with caps

2–3 empty aluminum soda cans

2 sheets of newspaper (for demonstration)

8 plastic garbage bags (5-gallon size)

8 twist ties

Rock salt

Vaseline® (optional, to seal stopper)

Marshmallows

Perishable Items

2–3 hard-boiled eggs (for Lesson III-4)

Nitrogen gas or air (enough for 1 balloon) (for Lesson IV-2)

1 helium balloon (for Lesson IV-2)

Carbon dioxide gas, from dry ice or the reaction of
baking soda and vinegar (for Lesson IV-2)

Ice (for Lessons I-1, I-4, II-2, II-4)

5–7 pounds dry ice (for Lesson I-4)

LESSON-BY-LESSON MATERIALS GUIDE

The kit contains the specialized items needed for each lab. However, common lab ware, chemicals, and other easily obtainable materials have not been included in this kit. Kit materials and other needed materials are listed separately below.

For consumable items, one kit contains enough materials for 10 classes of 32 students. Please note that Bunsen burners can be substituted for hot plates, but hot plates are safer to use.

Lesson I-1: Weather or Not

Preparation: Students work in pairs. Each pair needs an Erlenmeyer flask and a balloon. Other equipment can be shared between two pairs of students.

Materials Provided in Kit	Other Materials Needed
16 party balloons	32 pairs of safety goggles
16 oven mitts	16 250-mL Erlenmeyer flasks
	8–16 25-mL graduated cylinders
	8 hot plates
	8–16 buckets or large beakers of ice
	tap water

Lesson I-2: Raindrops Keep Falling . . .

Preparation: Students work in pairs. Each pair needs a graduated cylinder and a ruler. Other equipment can be shared between two pairs of students.

Materials Provided in Kit	Other Materials Needed
8 squirt bottles	16 100-mL beakers
16 12-in. transparent rulers	16 25-mL graduated cylinders
	plastic pipettes or droppers (optional)

Lesson I-3: Having a Melt Down

Materials Provided in Kit	Other Materials Needed
8 squirt bottles	8 25-mL graduated cylinders
	scales
	tap water

Lesson I-4: It's Sublime

Preparation: Students work in teams of four. Five pounds of dry ice is more than enough for five classes; check your phone directory for dry ice vendors. The hammer is needed to break the dry ice into smaller pieces. Students will use the 5-gallon cylindrical containers to measure the volume of the carbon dioxide gas.

Materials Provided in Kit	Other Materials Needed
8 oven mitts	32 pairs of safety goggles
	5–7 pounds of dry ice
	1 Styrofoam cooler or ice chest
	1 hammer
	8 plastic garbage bags (5-gallon size)
	8 twist ties
	scales
	2–4 5-gallon cylindrical containers
	Several ice cubes (for demonstration)
	2 400-mL beakers (for demonstration)
	1 hot plate (for demonstration)
	4–6 tongs (optional)

Lesson II-1: Hot Enough

Preparation: Students work in pairs. Each pair will need a small glass vial, a tightly fitting rubber stopper with a hole (or a septum with a hole), and a straw that fits tightly into the hole in the stopper. They can share hot plate, markers, and beakers of boiling water, room temperature water, and ice water with another pair.

Materials Provided in Kit	Other Materials Needed
16 small glass vials (2- or 4-dram size)	32 pairs of safety goggles
16 rubber stoppers with a hole	24 250-mL beakers
16 clear plastic straws	8 hot plates
food coloring	ice cubes
16 12-in. transparent rulers (metric)	Vaseline® (optional, to seal stopper)
16 permanent markers	

Lesson II-2: Full of Hot Air

Preparation: Students work in pairs. Each pair will need a test tube, a test tube holder, a capillary tube, a thermometer, two small rubber bands, and a permanent marker. They can share hot plate, beakers of boiling water, and beakers with an ice water/rock salt mixture with another pair.

Materials Provided in Kit	Other Materials Needed
16 1.5-mm capillary tubes closed on one end	32 pairs of safety goggles
32 small rubber bands	8 hot plates
food coloring	16 test tubes
16 12-in. transparent rulers (metric)	16 test tube holders
16 permanent markers	16 thermometers (Celsius scale)
few mL mineral oil	8 small beakers or vials (for oil)
	16 250-mL beakers
	8 hot plates
	tap water
	ice
	rock salt

Lesson II-4: It's Only a Phase

Preparation: Students work in pairs. Each pair will need a beaker, a thermometer, and an ice cube. If at all possible, the ice cube should be below 0°C (removed directly from a freezer before use). They can share hot plate with another pair.

Materials Provided in Kit	Other Materials Needed
16 oven mitts	32 pairs of safety goggles
	6 250-mL beaker
	16 ice cubes (kept below 0°C)
	16 thermometers
	8 hot plates
	ring stand and clamp to hold the thermometer (optional)

Lesson II-6: Hot Cement

Materials Provided in Kit	Other Materials Needed (for demonstration)
50 g potting soil (for demonstration)	desk lamp with high-intensity bulb (for demo)
2 clear plastic cups (for demonstration)	tap water
	2 thermometers (for demonstration)

Lesson III-1: Balancing Act

Materials Provided in Kit	Other Materials Needed
12 party balloons	2–3 pairs of safety goggles
2 50-mL syringes	1 2-L plastic bottle
3 ft of clear tubing	1 hot plate
1 clear plastic cup	2–3 empty aluminum soda cans
1 laminated card to fit over mouth of plastic cup	1 large shallow tub to catch spills
	tap water
	1 pair of tongs
	vacuum pump (recommended)
	vacuum desiccator or a side-arm flask with a stopper
	flasks, stoppers, glass tubing, and rubber tubing if vacuum pump is unavailable
	marshmallows (miniature if using syringe)

Lesson III-2: Feeling Under Pressure

Preparation: Pairs of students need a 50-mL syringe (without caps). The bathroom scale and the syringe that is capped will be used for a demonstration. If more bathroom scales are available, then more students can participate.

Materials Provided in Kit	Other Materials Needed
2 party balloons (for demonstration)	32 pairs of safety goggles
2 cups of sand (for demonstration)	1 funnel
16 50-mL syringes (without caps)	bathroom scale (one mandatory, more if available)
2 50-mL syringes with screw caps	

Lesson III-4: Egg in a Bottle

Materials Provided in Kit	Other Materials Needed
1 oven mitt	1 1000-mL Erlenmeyer flask (for demonstration)
	2–3 hard-boiled eggs, shelled (for demonstration)
	1 hot plate (for demonstration)

Lesson IV-1: Tower of Air

Materials Provided in Kit	Other Materials Needed
none	2 sheets of newspaper (for demonstration)
	1 wooden or stiff plastic ruler (for demonstration)

Lesson IV-2: Lighter Than Air

Materials Provided in Kit	Other Materials Needed
3 party balloons (for demonstration)	nitrogen gas or air (for demonstration)
	helium gas (for demonstration)
	carbon dioxide gas—can be generated from the reaction of baking soda and vinegar (for demonstration)

Lesson IV-3: More Than a Trillion

Materials Provided in Kit	Other Materials Needed
none	16 sets of 24 small pieces of paper—index cards or small Post-its® work well, or students can tear regular sheets of paper into 24 pieces

Lesson IV-4: Take a Breath

Preparation: Students work in teams of four. Each team needs a 2-L plastic bottle and a 3-ft piece of flexible tubing. Each student needs his or her own straw (or piece of a straw) to use as a clean mouthpiece.

Materials Provided in Kit	Other Materials Needed
8 3-ft pieces of flexible tubing	8 2-L plastic soda bottles with caps
32 short straws to fit snugly in tubing (straws can be cut)	8 graduated cylinders (preferably 250-mL or 500-mL)
8 wet-erase markers	tap water
	8 containers for water (~5-L size)

Lesson IV-5: Up in the Clouds

Preparation: Students can work in pairs or teams of four depending on how many plastic soda bottles are available. The dry 2-L bottle can be shared among the whole class or each team can have their own.

Materials Provided in Kit	Other Materials Needed
long matches	8–16 2-L plastic soda bottles with caps
	dry 2-L plastic soda bottle with cap
	tap water
	hot water (~80°C)

Lesson IV-6: Rain in the Forecast

Materials Provided in Kit	Other Materials Needed
50-mL rubbing alcohol	tap water
8–10 plastic pipettes	8–10 droppers (optional)

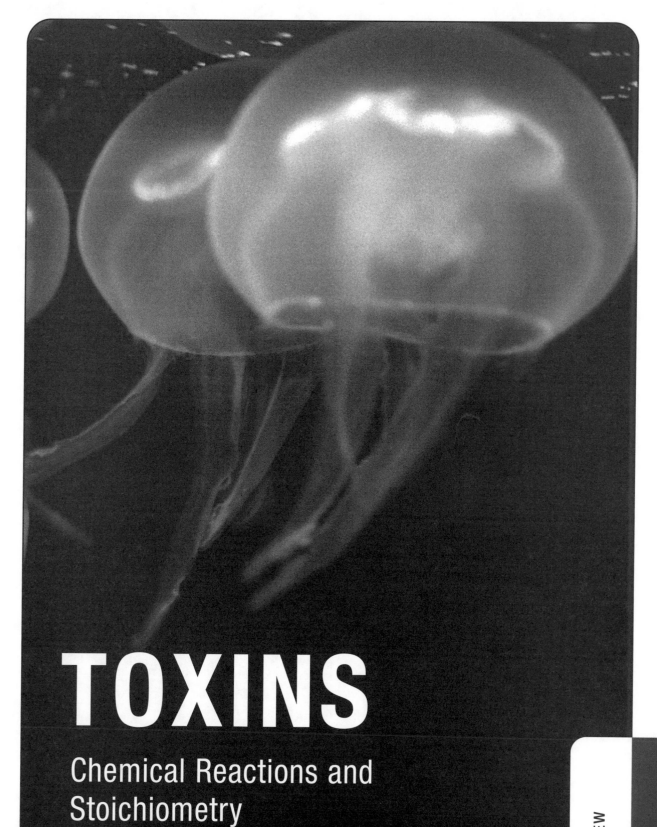

TOXINS

Chemical Reactions and
Stoichiometry

Contents

Main Topics Covered

COVERED IN DEPTH

Balancing chemical equations

Mole/number conversions

Mass/mole conversions

Relating masses of reactants and products

Properties of acids, bases, and salt solutions

Acids are hydrogen-ion-donating substances and bases are hydrogen-ion-accepting substances

pH scale

Definitions of acids and bases

Relationship between pH and hydrogen-ion concentration

Solutes and solvents

Dissolving process at the molecular level

Concentration

PREREQUISITES FOR THE UNIT

Periodic table

Atomic number

Atomic mass

Investigation Summaries

The study of toxins teaches students about solutions, concentrations, and chemical reactions. Within the framework of toxicity, the class learns about molarity, stoichiometry, and acids and bases.

INVESTIGATION I: DISSOLVING TOXINS

The study of toxicity begins with the idea that when taken in a great enough quantity, most substances are toxic and that toxicity is calculated in terms of its lethal dosage. This investigation explores solutions and teaches students about molarity and the effects of toxins on the human body. At the end of the investigation, students are able to identify substances by using information about their toxicity.

INVESTIGATION II: TRACKING TOXINS

This investigation explores chemical reactions. Toxins often do their damage through single-displacement or double-displacement reactions in the body. By conducting experiments, students observe the relationship between chemical formulas and their tangible versions, and identifies chemical and physical changes. They also learn about conservation of matter and how to balance chemical equations.

INVESTIGATION III: PRECIPITATING TOXINS

This investigation addresses precipitation reactions. Students learn to predict some reactions that will produce precipitates, and make both bones and kidney stones in the lab. They also explore stoichiometry and limiting reagents, ending with practice via a card game.

INVESTIGATION IV: NEUTRALIZING TOXINS

Toxins ends with a study of acids and bases, including pH and the Brønsted-Lowry definition. Students investigate reactions between acids and bases and learn about titrations.

BEFORE CLASS...

LESSON 1 – The Language of Change

Key Ideas:
A chemical equation is a chemical "sentence" describing either a
chemical change or a physical change. A chemical equation allows chemists to keep
track of the substances involved in a laboratory procedure. The reactants are the
substances that come together in a chemical reaction. The products are the substances
that form after a chemical reaction.

What Takes Place:
Students predict the outcome of a laboratory procedure based on the chemical
equation describing it. They then perform the procedure and keep careful track of
their observations, checking the outcome with their predictions. A final heating step
after the reaction allows students to check for the presence of sodium chloride in the
solution that was produced.

Materials: (per class of 32 students)
- Student worksheet
- 16 large test tubes
- 16 test tube holders (or 16 beakers to hold the test tube)
- 16 Bunsen burners (or alcohol burner)
- 8 labeled bottles of 3.0 M HCl (hydrochloric acid) – about 25 mL in each
- 8 labeled bottles of 3.0 M $NaHCO_3$ (sodium bicarbonate) – about 25 mL in each
- 32 safety goggles

**Safety note – Hydrochloric acid is dangerous and causes burns. Do not get
hydrochloric acid on your skin. In case of a spill, rinse with large amounts of
water. Wear goggles.**

Toxins © UC Regents, LHS Living by Chemistry, 2004.

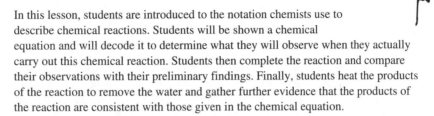

Investigation II – Tracking Toxins
LESSON 1 – The Language of Change

In this lesson, students are introduced to the notation chemists use to describe chemical reactions. Students will be shown a chemical equation and will decode it to determine what they will observe when they actually carry out this chemical reaction. Students then complete the reaction and compare their observations with their preliminary findings. Finally, students heat the products of the reaction to remove the water and gather further evidence that the products of the reaction are consistent with those given in the chemical equation.

Exploring the Topic   (5–10 min)

1. Introduce the ChemCatalyst exercise.

Write the following exercise on the board for students to complete individually.

Below is a chemical "sentence" describing the formation of a very toxic substance, hydrogen cyanide.

$$NaCN \text{ (s)} + HCl \text{ (aq)} \rightarrow NaCl \text{ (aq)} + HCN \text{ (g)}$$

- What kinds of information does this chemical notation contain? List at least four pieces of information contained in this chemical notation.

2. Discuss the ChemCatalyst exercise.

Use the discussion to get a sense of students' initial ideas.

Discussion goals:
Allow students to express their ideas about the information contained in a chemical equation.

Sample questions:
What information is contained in the above chemical "sentence"?
Why is there an arrow?
What is the difference between the left side and the right side?
What kinds of things do you think you would observe if you carried out this chemical reaction?

3. Introduce the terms reactant, product, and chemical equation.

Points to cover:
The chemical "sentence" above is called a **chemical equation.** A chemical equation describes change. The substances you start with before the change takes place are on the left side of the arrow. If there is more than one

TEACHER GUIDE

substance and they react with one another, they are called **reactants.** The substances you end up with after the change takes place are on the right side of the arrow. They are often referred to as **products**. This lesson will provide you with the opportunity to perform a chemical reaction and observe the change firsthand.

4. Explain the purpose of the activity.

Points to cover:

Tell students they will be making predictions about what they will see based on a chemical equation. Then they will carry out the reaction described in the equation. They will be gathering information to answer the question: "How are chemical reactions described using chemical notation?"

Activity – The Language of Change   (15 min)

5. Introduce the activity. (Worksheet)

Pass out the student worksheet for the lab. This is a two-part activity. Go over the procedure for Part II on the worksheet together. Divide the class into groups or pairs to do the lab.

Safety note – Hydrochloric acid is dangerous and causes burns. Do not get hydrochloric acid on your skin. In case of a spill, rinse with large amounts of water. Wear goggles.

Part I – Predict the outcome

The chemical notation and verbal description below both describe the same reaction. Compare the symbolic and verbal representations, and fill in the table that follows.

Chemical equation: $HCl(aq) + NaHCO_3(aq) \rightarrow NaCl(aq) + H_2O(l) + CO_2(g)$

Verbal description:

A solution of hydrochloric acid reacts with a solution of sodium bicarbonate to produce a solution of sodium chloride, water, and bubbles of carbon dioxide gas.

Symbol	What it Represents
HCl	hydrochloric acid
(aq)	aqueous
+	added to
$NaHCO_3$	sodium bicarbonate
→	reaction happens

Symbol	What it Represents
NaCl	sodium chloride
H_2O	water
(l)	liquid
CO_2	carbon dioxide
(g)	gas

86   Toxins © UC Regents, LHS Living by Chemistry, 2004.

Put an "R" next to the reactants in this chemical reaction. Put a "P" next to the products.

Part II – Perform the chemical reaction

Materials (for each pair of students)
1 large test tube
1 test tube holder
1 beaker to hold the test tube
1 Bunsen burner
1 10-mL graduated cylinder (can be shared with another pair of students)
1 labeled bottle of about 25 mL of 3.0 M HCl (hydrochloric acid) (to share with other pairs of students)
1 labeled bottle of about 25 mL of 3.0 M NaHCO$_3$ (sodium bicarbonate) (to share with other pairs of students)
2 safety goggles

Procedure:
1. Put on goggles.
2. Measure 2 mL of 3.0 M sodium bicarbonate (NaHCO$_3$), and pour it into the test tube.
3. Measure 2 mL of 3.0 M HCl, and slowly add to the test tube. Watch and listen to the reaction.
4. Gently shake the test tube after adding the 3.0 M HCl.
5. Record your observations in the data table below under **Observations During Reaction.**
6. Place your test tube into a test tube holder.
7. Describe the contents of the test tube in the data table below under **Observations After Reaction.**
8. Gently heat the test tube over the Bunsen burner (or alcohol burner). Make sure that you do not point the end of the test tube towards anyone. Be careful that the liquid does not splash out of the test tube.
9. Record your observations in the data table below under **Observations During Heating.**
10. Heat the test tube until all of the liquid is removed from the test tube, including any drops that are on the sides.
11. Describe the contents of the test tube in the data table below under **Observations After Heating.**
12. Clean up.

TEACHER GUIDE

Observations During Reaction	Possible observations: Bubbles in the liquid
Observations After Reaction	Clear solution
Observations During Heating	Bubbles, evaporation
Observations After Heating	Solid left in test tube

Answer the following questions:

1. What evidence do you have that CO_2 (g) was formed? (bubbles formed when the HCl(aq) was added)

2. What evidence do you have that NaCl (aq) was formed? (when the water evaporated, a solid remained)

3. What evidence do you have that H_2O (l) was formed? (since the reaction was carried out in water, it is difficult to tell if water was produced)

4. What did your observations tell you that the equation did not? (Observations tell you about the appearance of the substances – that the solutions are clear and colorless, that bubbles appear during the reaction.)

5. What does the equation tell you that your observations do not? (It tells you that a salt is present in the solution even though you cannot see it. It tells you the identity of the gas that was released.)

6. What was the purpose of heating the liquid after the reaction? (to reveal the solid salt that was dissolved in the water)

7. If you saw a solid powder settling to the bottom of a liquid at the end of the reaction, could the reaction be the one described above? Why or why not?

Making Sense:
Describe how a chemical equation keeps track of a chemical reaction.

Making Sense Discussion   (15 min)

Major goals: The main goal of this discussion is to make sure students can accurately translate the chemical notations in a chemical equation. They should also be able to relate the various parts of the equation to their real world experience of completing the reaction in the laboratory.

6. Relate the observations to the chemical equation.

Let students briefly share their observations from the lab. Encourage them to focus just on observations at this point, rather than on interpretations.

TEACHER GUIDE

Discussion goals:

Assist students in integrating their observations of a chemical reaction with their understanding of the corresponding chemical equation.

Sample questions:

What did you see and/or hear?

What observable changes took place at each stage in the procedure?

What evidence do you have that NaCl (aq) was formed?

What evidence do you have that H_2O (l) was formed?

What evidence do you have that CO_2 (g) was formed?

What did heating the products of the reaction reveal about the composition of those products?

What did your observations tell you that the equation didn't tell you?

What does the equation tell you that your observations didn't tell you?

Points to cover:

A chemical equation can help us to anticipate what change will occur. The equation can also tell us things that the reaction itself doesn't tell us. For instance, we do not know simply from observing the reaction between aqueous hydrochloric acid and aqueous sodium bicarbonate that a salt is dissolved in the clear liquid that has been produced. The equation gives us this information, and we can find evidence consistent with what the equation tells us by boiling away the liquid and seeing the remaining solid. This evidence strengthens our belief that the equation and the actual reaction match.

Students—like everyone else—may find it difficult to distinguish between their own perceptions and their interpretations of those perceptions. Help them to recognize when they are describing their observations and when they are going beyond their observations—drawing inferences, imposing assumptions, applying information and ideas they have acquired in other contexts.

7. Breakdown of a chemical equation.

You may wish to write the equation below on the board and refer to the different parts during the discussion.

Discussion goals:

Focus the discussion on the information contained in a chemical equation.

Sample questions:

What things does a chemical equation keep track of?

Points to cover:

A chemical equation consists of chemical formulas for the compound or elements that are undergoing some sort of change. Each chemical formula can be translated into a specific substance. The items on the left of the arrow indicate the substances before any change has taken place. The items on the right side of the arrow indicate what substances are produced after the change.

TEACHER GUIDE

The arrow essentially represents "something happening" or "a change". A chemical equation also indicates what form the substances are in at all stages – solid, liquid, gas, or aqueous.

$$HCl(aq) + NaHCO_3(aq) \rightarrow NaCl(aq) + H_2O(l) + CO_2(g)$$

We will learn more details about chemical equations in later lessons.

Check-in   (5 min)

8. Introduce the Check-in exercise.

Write the following exercise on the board for students to complete individually.

Consider the following reaction between sodium cyanide and a solution of hydrochloric acid.

$$NaCN\,(s) + HCl\,(aq) \rightarrow NaCl\,(aq) + HCN\,(g)$$

- Describe in detail what you would observe if you carried out this reaction.
- Describe the products that you would have.

9. Discuss the Check-in exercise.

Get a sense of the level of understanding by taking a vote, collecting students' work, or asking students to defend their choices.

Discussion goals:
Make sure students are connecting chemical equations to actual observations.

Sample questions:
    What would you see at the beginning and end of the reaction?
    How might poisoning occur if you carried out this reaction?

Students should realize that they begin with a solid and a clear liquid. The products are a clear liquid and a toxic gas. Therefore, it is likely that they will be poisoned by inhalation. However, if they ingest the NaCN (s), it would be toxic because it will produce HCN gas when it reacts with stomach acid.

10. Wrap-up.

Assist students in summarizing what was learned in this class.
- Chemical equations help chemists keep track of the substances involved in chemical and physical changes.
- Chemical equations indicate the reactants and products of chemical reactions.

Homework

11. Assign homework.

Use the homework provided with the curriculum or assign your own.

90   Toxins © UC Regents, LHS Living by Chemistry, 2004.

Homework – Investigation II – Lesson 1

Write a lab report for today's class.

TEACHER GUIDE

Student Worksheet

The Language of Change

Name: _____

Period: _____ Date: _____

Purpose: In this experiment, you will carry out the reaction between hydrochloric acid and sodium bicarbonate, and then do some analysis of the products that form.

Part I – Predict the outcome

The chemical notation and verbal description below both describe the same reaction. Compare the symbolic and verbal representations, and fill in the table that follows.

Chemical equation: $HCl(aq) + NaHCO_3(aq) \rightarrow NaCl(aq) + H_2O(l) + CO_2(g)$

Verbal description:

A solution of hydrochloric acid reacts with a solution of sodium bicarbonate to produce a solution of sodium chloride, water, and bubbles of carbon dioxide gas.

Symbol	What It Represents
HCl	
(aq)	
+	
NaHCO$_3$	
→	

Symbol	What It Represents
NaCl	
H$_2$O	
(l)	
CO$_2$	
(g)	

Put an "R" next to the reactants in this chemical reaction. Put a "P" next to the products.

Part II – Perform the chemical reaction

Procedure

1. Put on your goggles.
2. Measure 2 mL of 3.0 M sodium bicarbonate (NaHCO$_3$), and add it into the test tube.
3. Measure 2 mL of 3.0 M HCl, and add it slowly to the test tube. Watch and listen to the reaction.
4. Gently shake the test tube after adding the 3.0 M HCl.
5. Record your observations in the data table below under **Observations During Reaction.**
6. Place your test tube into a test tube holder.
7. Describe the contents of the test tube in the data table below under **Observations After Reaction.**
8. Gently heat the test tube over the Bunsen burner (or alcohol burner). Make sure that you do not point the end of the test tube towards anyone. Be careful that the liquid does not splash out of the test tube.
9. Record your observations in the data table below under **Observations During Heating.**
10. Heat the test tube until all of the liquid is removed from the test tube, including any drops that are on the sides.

11. Describe the contents of the test tube in the data table below under **Observations After Heating.**
12. Clean up.

Data

Observations During Reaction	
Observations After Reaction	
Observations During Heating	
Observations After Heating	

$$HCl(aq) + NaHCO_3(aq) \rightarrow NaCl(aq) + H_2O(l) + CO_2(g)$$

1. What evidence do you have that CO_2 (g) was formed?

2. What evidence do you have that NaCl (aq) was formed?

3. What evidence do you have that H_2O (l) was formed?

4. What did your observations tell you that the equation did not?

5. What does the equation tell you that your observations did not?

6. What was the purpose of heating the liquid after the reaction?

7. If you saw a solid powder settling to the bottom of a liquid at the end of the reaction, could the reaction be the one described above? Why or why not?

Making Sense:
Describe how a chemical equation keeps track of a chemical reaction.

TEACHER GUIDE

Investigation II – Tracking Toxins

A chemical equation is a chemical "sentence" describing either a chemical change or a physical change. A chemical equation allows chemists to keep track of the substances involved in a laboratory procedure or a chemical reaction. This investigation introduces you to chemical reactions and the chemical equations that describe them.

Lesson Guide: The Language of Change

Investigation II – Lesson 1

This activity allows you to predict the outcome of a chemical equation and then test your predictions by completing the reaction experimentally.

ChemCatalyst

Answer the following question:

Below is a chemical "sentence" describing the formation of a very toxic substance, hydrogen cyanide:

$$NaCN\ (s) + HCl\ (aq) \rightarrow NaCl\ (aq) + HCN\ (g)$$

- What kinds of information does this chemical notation contain? List at least four pieces of information contained in this chemical notation.

Activity

- ❖ Safety note – Hydrochloric acid is dangerous and causes burns. Do not get hydrochloric acid on your skin. In case of a spill, rinse with large amounts of water. Wear goggles.

Part I – Predict the outcome
The chemical notation and verbal description below both describe the same reaction. Compare the symbolic and verbal representations, and fill in the table that follows.

Chemical equation: $HCl(aq) + NaHCO_3(aq) \rightarrow NaCl(aq) + H_2O(l) + CO_2(g)$

Verbal description:
A solution of hydrochloric acid reacts with a solution of sodium bicarbonate to produce a solution of sodium chloride, water, and bubbles of carbon dioxide gas.

Toxins © UC Regents, LHS Living by Chemistry, 2004.   35

Copy this table in your notebook

Symbol	What it Represents
HCl	
(aq)	
+	
NaHCO$_3$	
→	

Symbol	What it Represents
NaCl	
H$_2$O	
(l)	
CO$_2$	
(g)	

Put an "R" next to the reactants in this chemical reaction. Put a "P" next to the products.

Part II – Perform the chemical reaction

Procedure:

1. Before you begin, create a date table similar to the one below, for your observations.

2. Put on goggles.

3. Measure 2 mL of 3.0 M sodium bicarbonate (NaHCO$_3$), and pour it into the test tube.

4. Measure 2 mL of 3.0 M HCl, and slowly add to the test tube. Watch and listen to the reaction.

5. Gently shake the test tube after adding the 3.0 M HCl.

6. Record your observations in your data table under Observations During Reaction.

7. Place your test tube into a test tube holder.

8. Describe the contents of the test tube in your data table under Observations After Reaction.

9. Gently heat the test tube over the Bunsen burner (or alcohol burner). Make sure that you do not point the end of the test tube towards anyone. Be careful that the liquid does not splash out of the test tube.

10. Record your observations in your data table under Observations During Heating.

11. Heat the test tube until all of the liquid is removed from the test tube, including any drops that are on the sides.

12. Describe the contents of the test tube in your data table under Observations After Heating.

13. Clean up.

Toxins© UC Regents, LHS Living by Chemistry, 2004.

STUDENT GUIDE

Data Table:

Observations During Reaction	
Observations After Reaction	
Observations During Heating	
Observations After Heating	

Copy this table in your notebook

$$HCl(aq) + NaHCO_3(aq) \rightarrow NaCl(aq) + H_2O(l) + CO_2(g)$$

Answer the following questions:

1. What evidence do you have that CO_2 (g) was formed?

2. What evidence do you have that NaCl (aq) was formed?

3. What evidence do you have that H_2O (l) was formed?

4. What did your observations tell you that the equation did not?

5. What does the equation tell you that your observations do not?

6. What was the purpose of heating the liquid after the reaction?

7. If you saw a solid powder settling to the bottom of a liquid at the end of the reaction, could the reaction be the one described above? Why or why not?

Making Sense Question:
Describe how a chemical equation keeps track of a chemical reaction.

Check-in

Answer the following questions:

Consider the following reaction between sodium cyanide and a solution of hydrochloric acid.

$$NaCN\ (s) + HCl\ (aq)\ \rightarrow\ NaCl\ (aq) + HCN\ (g)$$

- Describe in detail what you would observe if you carried out this reaction.
- Describe the products that you would have.

STUDENT GUIDE

Homework

Complete the following for homework:

Write a complete lab report.

Materials for *Toxins* Unit

KIT CONTENTS

16 sets of colored blocks (Each set contains 50 white, 50 yellow, 100 blue, 200 purple. Exact colors may vary.)

32 plastic cups[R]

48 plastic spoons[R]

1 bag gummy bears[R]

184 small dropper bottles

56 medium dropper bottles

10 large dropper bottles

labels for all bottles[R]

10 packets unsweetened Kool-Aid®

1/4 cup sand

1 canister sweetened Kool-Aid®

8 wet-erase markers[R]

1 marking pen

3 cups rice[R]

3 cups lentils[R]

200 plastic baggies

64 plastic transfer pipettes

16 24-well plates

12 clear vials

8 Proton Shuffle card sets

8 Grammies card sets

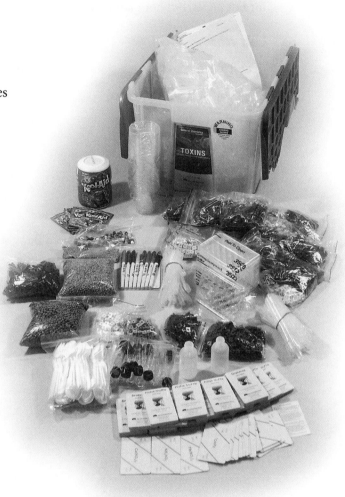

For up-to-date information about kit contents, please visit www.keypress.com/chemistry .

[R]Consumable; will need to be replenished. This kit includes enough materials for approximately 10 classes of 32 students, or 2–3 years.

OTHER MATERIALS NEEDED

Common lab ware, supplies, chemicals, and other easily obtainable materials have not been included in this kit. They are listed below.

Teachers should follow all state, local, and district guidelines concerning laboratory safety practices, safe handling of materials, and disposal of hazardous materials. In the absence of guidelines related to the handling, storage, and disposal of laboratory materials, teachers should refer to the Materials Safety Data Sheet (MSDS) for each chemical.

Your classroom should have a sink. Please note that Bunsen burners can be substituted for hot plates, but hot plates are safer to use.

Lab Ware

(These quantities are for a class of 32 students.)

32 pairs of safety goggles

8 scales

16 Bunsen burners

8 wash bottles

8 waste containers

16 50-mL beakers or small containers

32 250-mL beakers (plastic cups can be substituted)

6 400-mL beakers

16 500-mL beakers

3 1-L beakers (or plastic bottles)

8 50-mL graduated cylinders

8 100-mL graduated cylinders

40 medium test tubes

36 large test tubes (16 × 150 mm)

16 test tube racks

8 stir rods

8 transfer pipettes

6 containers to hold 5 g of solids: Film canisters or small beakers are fine.

weigh boats (preferred) (weigh paper or wax paper squares can be used)

9 small scoops or spatulas (or plastic straws flared at one end)

8 cork stoppers (optional—to put on test tube for shaking)

1 hammer

Chemicals

(These quantities are for 10 classes of 32 students.)

phenol red

phenolphthalein indicator

21 g calcium hydroxide $Ca(OH)_2$

55 g calcium chloride $CaCl_2$

50 g calcium carbonate (powdered or chunk; chalk is fine)

5 g ammonium hydroxide NH_4OH

9 g sodium hydroxide NaOH

10 g zinc (powdered or "mossy")

90 g copper sulfate $CuSO_4$

75 g potassium chloride KCl

110 g sodium bromide NaBr

78 g sodium hydroxide NaOH

100 mL ethanol

100 mL butanol

200 mL acetic acid

25 g potassium nitrate KNO_3

30 g magnesium nitrate $Mg(NO_3)_2$

40 g copper nitrate $Cu(NO_3)_2$

40 g silver nitrate $AgNO_3$

60 g sodium chloride NaCl

58 g sodium bicarbonate Na_2CO_3

150 g baking soda $NaHCO_3$

200 mL 1.0 M hydrochloric acid HCl

500 mL 3.0 M hydrochloric acid HCl

25 g sodium carbonate (washing soda)

10 g sodium oxalate $Na_2C_2O_4$

20 g sodium phosphate $Na_3PO_4 \cdot 12H_2O$

8 bottles of universal indicator with color key

Other Supplies

2 cups dry ice (for Lesson II-2)

small cooler to hold dry ice, labeled CO_2

250 g sucrose (refined sugar)

200 mL corn syrup

50 g kosher salt or lab-grade NaCl

1/4 cup Epsom salts or table salt

1 head red cabbage (for Lesson IV-1)

blender or pot to boil cabbage juice (for Lesson IV-1)

strainer for cabbage juice (for Lesson IV-1)

200 mL vinegar

200 mL lemon juice

200 mL window cleaner with ammonia

200 mL rubbing alcohol

200 mL distilled water

tap water

100 mL oil (mineral oil, baby oil, or vegetable oil)

1/4-cup measuring cup

8 small paper cups

16 straws

8 water bottles

8–16 index cards

Glue (to attach gummy bears to index cards)

LESSON-BY-LESSON MATERIALS GUIDE

See Stock Solution Recipes on page xi of the *Toxins* Teacher Guide.

Lesson I-2: Bearly Alive

Preparation: The gummy bears need to soak in their solutions for at least two hours (preferably overnight). Four students share one set of gummy bears. The students can work in two pairs or as a team of four.

Materials Provided in Kit	Other Materials Needed
4 sets of labels (8 labels per set): water, 2.0 M sugar, 1.0 M sugar, 0.1 M sugar, corn syrup, 1.0 M salt, 0.5 M salt, 0.1 M salt	32 pairs of safety goggles
32 large plastic cups	8–16 index cards
32 plastic spoons	glue (to attach gummy bears to index cards)
80 gummy bears (red or green to soak, the other colors can be attached to the index cards)	250 g sucrose (refined sugar)
	tap water
	200 mL corn syrup
	50 g kosher salt or lab-grade NaCl

	STOCK SOLUTIONS:
	    200 mL of 2.0 M sugar solution
	    200 mL of 1.0 M sugar solution
	    200 mL of 0.1 M sugar solution
	    200 mL of 2.0 M salt solution
	    200 mL of 1.0 M salt solution
	    200 mL of 0.1 M salt solution

Lesson I-3: Mixing It Up!

Preparation: Students work in teams of four. Two teams share the ethanol, butanol, oil, phenol red, water, and copper sulfate. If you don't have small scoops or spatulas, you can use straws cut at one end to a 45-degree angle to create a scoop.

Materials Provided in Kit	Other Materials Needed
8 small dropper bottles	3 400-mL beakers
16 medium dropper bottles	40 medium test tubes
5 sets of labels (5 labels per set): phenol red, ethanol, butanol, water, and oil	8 test tube racks
	3 stir rods
1 packet unsweetened Kool-Aid®	tap water
1/4 cup sand	1/4 cup Epsom salts or table salt
8 marking pens (for labeling test tubes)	100 mL ethanol
	100 mL butanol
	100 mL oil (mineral oil, baby oil, or vegetable oil)
	phenol red indicator
	40 g copper sulfate
	16 straws
	8 small scoops or spatulas (or plastic straws flared at one end)
	8 cork stoppers (optional—to put on test tube for shaking)
	STOCK SOLUTION: 100 mL of 0.1% phenol red solution

Lesson I-4: Weighing In

Preparation: Students work in teams of four.

Materials Provided in Kit	Other Materials Needed
3 cups rice	8 scales
3 cups lentils	periodic tables
16 plastic baggies	8 small paper cups
	1/4-cup measuring cup

Lesson I-5: Finding Solutions

Preparation: Students work in teams of four.

Materials Provided in Kit	Other Materials Needed
1 canister powdered sweetened Kool-Aid®	weigh boats (preferred) (weigh paper or wax paper squares can be used)
16 plastic transfer pipettes	8 stirring rods or Popsicle® sticks
8 marking pens (for labeling beakers)	8 100-mL graduated cylinders
	8 water bottles
	32 250-mL beakers (plastic cups can be substituted)
	8 scales

Lesson I-7: Is It Toxic?

Materials Provided in Kit	Other Materials Needed
none	32 pairs of safety goggles
	8 50-mL graduated cylinders
	24 250-mL beakers
	tap water
	3 1-L beakers (or plastic bottles) labeled solution #1, solution #2, solution #3 (solution #1 is 1.0 M KCl, #2 is 1.0 M NaBr, #3 is 1.0 M NaOH)
	8 scales
	8 small paper cups
	8 transfer pipettes
	75 g potassium chloride
	110 g sodium bromide
	50 g sodium hydroxide
	8 pens to label cups
	STOCK SOLUTIONS: 1.0 L of 1.0 M KCl 1.0 L of 1.0 M NaBr 1.0 L of 1.0 M NaOH

Lesson II-1: The Language of Change

Preparation: Students work in pairs.

Materials Provided in Kit	Other Materials Needed
16 medium dropper bottles	32 pairs of safety goggles
8 sets of labels (2 labels per set): 3.0 M hydrochloric acid (HCl), 3.0 M sodium bicarbonate (NaHCO$_3$)	16 test tube holders
	16 large test tubes (16 × 150 mm)
	16 Bunsen burners
	test tube racks or small beakers to hold test tubes
	150 g NaHCO$_3$ (baking soda)
	8 10-mL graduated cylinders
	STOCK SOLUTIONS: 500 mL 3.0 M NaHCO$_3$ 500 mL of 3.0 M HCl

Lesson II-2: Making Predictions

Preparation: This lab is set up as a station activity with enough supplies for three student teams of four at each station.

Materials Provided in Kit	Other Materials Needed
10 large dropper bottles	32 pairs of safety goggles
6 sets of labels (5 labels per set): saturated calcium hydroxide Ca(OH)$_2$, 1.0 M CaCl$_2$, 1.0 M NH$_4$OH, 1.0 M copper sulfate CuSO$_4$, 1.0 M sodium hydroxide (NaOH)	6 wash bottles (2 per station for rinsing test tubes)
	6 labeled water bottles for the lab station experiments
	36 test tubes (16 × 150 mm)
	6 test tube racks (or beakers to hold test tubes)
	6 400-mL beakers for disposing contents of test tubes labeled Waste 1–9
	2 cups dry ice
	1 hammer (to smash dry ice)
	small cooler to hold dry ice labeled CO$_2$ (s)
	9 small spatulas
	1 g Ca(OH)$_2$
	25 g calcium chloride CaCl$_2$

(continued)

Lesson II-2: Making Predictions (continued)

Materials Provided in Kit	Other Materials Needed
	50 g copper sulfate $CuSO_4$
	5 g NH_4OH
	5 g NaOH
	10 g Zn (powdered or "mossy")
	6 small containers for the solids (2 labeled $CaCl_2$ (s), 2 labeled $CuSO_4$ (s), and 2 labeled Zn(s)—put approximately 5 g of solid in each. Film canisters or small beakers are fine.

	STOCK SOLUTIONS:
	    100 mL of 1.0 M $CaCl_2$
	    100 mL of 1.0 M NH_4OH
	    100 mL of 1.0 NaOH
	    100 mL of 1.0 M $CuSO_4$
	    100 mL of saturated $Ca(OH)_2$

Lesson II-5: Some Things Never Change

Preparation: Students work in teams of four. If you do the reactions as a demonstration, use only about 100 mL of each solution.

Materials Provided in Kit	Other Materials Needed
32 large plastic cups	32 pairs of safety goggles
8 sets of labels: 1.0 M Na_2CO_3, 1.0 M $CaCl_2$, acetic acid ($C_2H_4O_2$)	8 scales
8 marking pens	24 100–250-mL flasks or beakers to hold student stock solutions labeled 1.0 M Na_2CO_3, 1.0 M $CaCl_2$, and acetic acid $C_2H_4O_2$ (vinegar)
	25 g $CaCl_2$
	50 g Na_2CO_3
	200 mL acetic acid

	STOCK SOLUTIONS:
	    200 mL of 1.0 M $CaCl_2$
	    400 mL of 1.0 M Na_2CO_3

Lesson II-6: Atom Inventory

Materials Provided in Kit	Other Materials Needed
16 sets of colored blocks (each set contains 50 white, 50 yellow, 100 blue, 200 purple. Exact block colors may vary.): 1 set per student pair	none

Lesson III-1: Solid Evidence

Disposal: Students are testing heavy metals. Be sure to dispose of chemicals properly. Check state and district guidelines, or refer to the Materials Safety Data Sheet for each chemical.

Preparation: The silver nitrate solution needs to be stored in an opaque container. If you plan to store the solutions in their dropper bottles, make sure the silver nitrate is in an opaque bag or box. Each pair of students will use their own well plate, but two pairs (four students) will share the dropper bottles of solution.

Materials Provided in Kit	Other Materials Needed
16 24-well plates	32 pairs of safety goggles
8 sets of labels: 1.0 M KNO_3, 1.0 M $Mg(NO_3)_2$, 1.0 M $Cu(NO_3)_2$, 1.0 M $AgNO_3$, 1.0 M NaCl, 1.0 M Na_2CO_3, and 1.0 M NaOH	8 wash bottles for rinsing plates
	8 plastic dump tubs or large beakers to collect waste for proper disposal
56 small dropper bottles	25 g KNO_3
	30 g $Mg(NO_3)_2$
	40 g $Cu(NO_3)_2$
	40 g $AgNO_3$
	15 g NaCl
	25 g Na_2CO_3
	8 g NaOH
	STOCK SOLUTIONS: 200 mL of 1.0 M KNO_3 200 mL of 1.0 M $Mg(NO_3)_2$ 200 mL of 1.0 M $Cu(NO_3)_2$ 200 mL of 1.0 M $AgNO_3$—**solution should be stored in an opaque container** 200 mL of 1.0 M NaCl 200 mL of 1.0 M Na_2CO_3 200 mL of 1.0 M NaOH

Lesson III-3: Sticks and Stones

Disposal: Students are testing heavy metals. Be sure to dispose of chemicals properly. Check state and district guidelines, or refer to the Materials Safety Data Sheet for each chemical.

Materials Provided in Kit	Other Materials Needed
16 well plates	10 g sodium oxalate $Na_2C_2O_4$
24 medium dropper bottles	20 g sodium phosphate $Na_3PO_4 \cdot 12H_2O$
8 sets of labels (4 labels per set): 0.1 M $CaCl_2$, 0.1 M sodium oxalate, 0.1 M sodium phosphate	5 g $CaCl_2$
	32 pairs of safety goggles
8 pens to label test tubes	8 test tube racks
	8 wash bottles
	8 waste containers
	STOCK SOLUTIONS: 200 mL of 0.10 M $CaCl_2$ 200 mL of 0.10 M $Na_2C_2O_4$ 200 mL of 0.10 M Na_3PO_4

Lesson III-4: Blockhead

Materials Provided in Kit	Other Materials Needed
16 sets of colored blocks (each set contains 50 white, 50 yellow, 100 blue, 200 purple. Exact block colors may vary.): 1 set per student pair	8 scales

Lesson III-7: Grammies

Materials Provided in Kit	Other Materials Needed
6 Grammies card sets (36 cards per set)	none

Lesson IV-1: Heartburn

Preparation: Make sure to allow yourself time to prepare the cabbage juice indicator. Allow one hour advance preparation if boiling the cabbage juice.

Students work in pairs. Two pairs share bottles of solution and indicator, wash bottles, and waste containers.

Materials Provided in Kit	Other Materials Needed
16 24-well plates	1 head of red cabbage
88 small dropper bottles	blender or pot to boil cabbage juice
8 sets of labels (10 labels per set): cabbage juice, universal indicator, salt solution (1.0 M NaCl), drain cleaner (0.1 M NaOH), stomach acid (0.1 M HCl), ammonium hydroxide (NH_4OH), distilled water (H_2O), rubbing alcohol (C_3H_9OH), washing soda (Na_2CO_3), vinegar ($C_2H_4O_2$), lemon juice ($C_6H_8O_7$)	strainer for cabbage juice
	8 wash bottles
	8 waste containers
	50 g calcium carbonate (powdered or chunk; chalk is fine)
	200 mL vinegar
	200 mL window cleaner with ammonia
	200 mL distilled water
	200 mL rubbing alcohol
	25 g sodium carbonate (washing soda)
	15 g sodium chloride
	200 mL lemon juice
	8 small spatulas
	8 20–50-mL beakers or small containers for calcium carbonate
	8 bottles of universal indicator with color key

	STOCK SOLUTIONS:
	    200 mL of 1.0 M NaCl
	    200 mL of 1.0 M Na_2CO_3
	    200 mL of 0.1 M NaOH
	    200 mL of 0.1 M HCl

Lesson IV-2: Watered Down

Preparation: The gummy bears need to be prepared the night before. Let them soak in solutions of 1.0 M HCl, 0.1 M HCl, and distilled water.

Students work in pairs. Two pairs share bottles of solution and indicator, wash bottles, and waste containers.

Materials Provided in Kit	Other Materials Needed
12 clear vials	distilled water
16 24-well plates	16 250-mL beakers
32 small dropper bottles	16 500-mL beakers
8 sets of labels (4 labels per set): 1.0 M HCl, 1.0 M NaOH, 1.0 M NaCl, universal indicator	8 bottles of universal indicator with color key
32 plastic transfer pipettes	8 wash bottles
8 pens to label beakers	8 waste containers
12 red or green gummy bears	20 g sodium hydroxide
	30 g sodium chloride
	200 mL of 1.0 M HCl

	STOCK SOLUTIONS:
	    200 mL of 1.0 M NaCl
	    200 mL of 1.0 M NaOH
	    200 mL of 0.1 M HCl

Lesson IV-4: Proton Shuffle

Materials Provided in Kit	Other Materials Needed
8 sets of Proton Shuffle cards (27 cards per set)	none

Lesson IV-5: Neutral Territory

Preparation: Students work in pairs. Each pair has their own well plate, but two pairs share the rest of the materials.

Materials Provided in Kit	Other Materials Needed
16 well plates	32 pairs of safety goggles
32 small dropper bottles	8 wash bottles
8 sets of labels (8 labels per set): 0.10 M HCl (hydrochloric acid), 0.10 M HNO$_3$ (nitric acid), 0.10 M NaOH (sodium hydroxide), universal indicator, and Ca(OH)$_2$ (calcium hydroxide)	8 waste containers
	20 g calcium hydroxide
	8 small containers for calcium hydroxide
	8 bottles of universal indicator with color key
	2 g sodium hydroxide

	STOCK SOLUTIONS:
	    200 mL of 0.1 M HCl
	    200 mL of 0.1 M HNO$_3$
	    200 mL of 0.1 M NaOH

Lesson IV-6: Drip Drop

Materials Provided in Kit	Other Materials Needed
40 small dropper bottles	32 pairs of safety goggles
8 sets of labels (5 labels per set): HCl Solution A, HCl Solution B, HCl Solution C, 0.10 M NaOH, and phenolphthalein indicator	8 wash bottles
	8 waste containers
	16 50-mL beakers
	phenolphthalein indicator
	2 g sodium hydroxide
	STOCK SOLUTIONS: 200 mL of 0.05 M HCl 200 mL of 0.1 M HCl 200 mL of 0.02 M HCl 200 mL of 0.1 M NaOH

FIRE

Energy and Thermochemistry

Contents

Main Topics Covered

COVERED IN DEPTH

Temperature and heat flow

Exothermic and endothermic processes

Energy associated with phase changes

Heat flow problems, specific heat capacity, latent heat

Enthalpy of reaction, Hess's law

Catalyst

Activation energy

PREREQUISITES FOR THE UNIT

Periodic table, atomic number, atomic mass

Chemical reactions

Investigation Summaries

The *Fire* unit focuses on energy in the context of the often spectacular phenomena we call fire. Students investigate the intimate connection between energy and changes in matter as they learn about heat transfer, calorimetry, combustion, activation energy, bond energies, and enthalpy changes.

INVESTIGATION I: EVIDENCE OF CHANGE

The first investigation of the *Fire* unit introduces students to the basic concepts and vocabulary associated with heat and heat transfer. Beginning with discussions of "hot" and "cold" in terms of heat transfer, and moving through endothermic and exothermic reactions and calorimetry, this investigation gives students a firm foundation from which to study combustion.

INVESTIGATION II: CONDITIONS FOR CHANGE

This investigation focuses on combustion: how to achieve it and how to sustain it. Students are first asked to consider the conditions necessary for a fire, beginning with the fire triangle and continuing as they discover the properties of effective fuels. The study of combustion and balanced chemical equations ends with a lab experiment where students make and test their own sparklers.

INVESTIGATION III: ENERGY FOR CHANGE

The third investigation of the *Fire* unit explores combustion on a particulate level, moving from the energy transfer inherent in the making and breaking of bonds through studies of activation energy, the law of conservation of energy, and enthalpy. Students summarize and reinforce what they have learned by writing combustion equations and creating a Fire Concept Map.

LESSON 3 – All-A-Glow

Key Ideas:

When carbon compounds are burned they produce carbon dioxide, water, and heat as products. Flames are the result of heated gases that are produced during a combustion reaction. These heated gases emit light. Combustion reactions that don't produce gases don't have flames. When metals combust, metal oxides and heat are produced. The metals and metal oxides are solids. A glow is observed on the solid as the combustion reaction proceeds, but there is no flame.

What Takes Place:

Students examine balanced chemical equations for a number of combustion reactions. They examine the combustion of carbon-containing molecules to look for patterns that depend on the composition of the fuel. Students also examine reactions of metals and metal salts with oxygen. The lesson ends with a discussion as to why a flame is observed for the combustion of carbon-containing molecules, whereas only a glow is observed for the combustion of metals.

Materials: (For each class)

- Student worksheet

Investigation II – Conditions for Change

LESSON 3 – All-A-Glow

In the previous two lessons, we learned that oxygen and fuels are needed for combustion. Both carbon-containing molecules and metals can serve as fuels. In this lesson, we will examine the products of combustion by writing balanced chemical equations. Students will look for patterns in the identities and amounts of reactants and products. The lesson ends with a discussion of flames, and the connection between flames and combustion reactions that occur in the gas phase.

Exploring the Topic   (5–10 min)

1. Introduce the ChemCatalyst exercise.

Write the following exercise on the board for students to complete individually.

The following table shows the balanced chemical equations for four combustion reactions.

Substance	Combustion reaction
methane	$CH_4 + 2\,O_2 \rightarrow CO_2 + 2\,H_2O$
ethanol	$C_2H_6O + 3\,O_2 \rightarrow 2\,CO_2 + 3\,H_2O$
glucose	$C_6H_{12}O_6 + 6\,O_2 \rightarrow 6\,CO_2 + 6\,H_2O$
magnesium	$2\,Mg + O_2 \rightarrow 2\,MgO$

- List three patterns you notice.

2. Discuss the ChemCatalyst exercise.

Use the discussion to get a sense of students' initial ideas.

<u>Discussion goals:</u>
Use the students' written responses to stimulate an open-ended discussion of chemical equations for combustion reactions.

Sample questions:
  Are the reactions balanced?
  Which substances in the combustion reactions are fuels?
  What are the products of combustion?
  What similarities do you notice for the combustion of the three carbon-containing molecules?
  How is the combustion of magnesium different from the other three?
  What do all four combustion reactions have in common?

3. Explore the balancing of combustion equations.

Write the skeleton equation below on the board for the combustion of ethane. Ask students to assist you with balancing equations.

$$C_2H_6 \;+\; O_2 \;\rightarrow\; CO_2 \;+\; H_2O$$

<u>Discussion goals:</u>

Assist students in understanding the pros and cons of different stoichiometric representations.

Sample questions:

Balance the equation on the board for the combustion of ethane.

$$2\,C_2H_6 \;+\; 7\,O_2 \;\rightarrow\; 4\,CO_2 \;+\; 6\,H_2O$$

What if we were interested in looking at only one mole of the fuel at a time, how would we change the equation? (Divide coefficients by 2)

$$C_2H_6 \;+\; 3.5\,O_2 \;\rightarrow\; 2\,CO_2 \;+\; 3\,H_2O$$

What is the ratio of ethane molecules to oxygen molecules that react? (2:7)

How many molecules of water can you make by combusting one molecule of ethane? Which equation did you use to answer the question?

What advantages are there to the two different representations?

Points to cover:

The first balanced equation allows us to look at all the compounds in terms of whole numbers. Thus, two molecules of ethane react with seven molecules of oxygen to produce four carbon dioxide molecules and six water molecules. The second balanced equation allows us to consider every compound in terms of one unit of ethane (whether it be one molecule or one mole). Often it is more useful to consider a stoichiometric equation in terms of one particular species – in this case, we are focusing on the fuel being combusted.

4. Explain the purpose of today's activity.

If you wish you can write the main question on the board.

Points to cover:

Tell students they will be examining chemical equations for a number of combustion reactions. They will examine patterns in the identities and amounts of reactants and products. The question for today is: "How can we write a chemical equation to describe a combustion reaction?"

Activity – All-A-Glow   (15 min)

5. Introduce the activity. (Worksheet)

Pass out the worksheet. Ask students to work individually.

TEACHER GUIDE

Answer the following questions:

1. Molecules containing carbon and hydrogen are called alkanes. Alkanes react with oxygen to produce carbon dioxide, water, and a fire.

   a. Draw the structural formulas for ethane (C_2H_6), and butane (C_4H_{10}).

   b. The reactions for methane, ethane, and hexane are balanced in the table. Balance the reactions for propane, butane, and pentane. (Solve for one unit of fuel in each case.)

Combustion Reactions of Alkanes		
methane	CH_4 + $2\,O_2$ → CO_2 + $2\,H_2O$	
ethane	C_2H_6 + $3.5\,O_2$ → $2\,CO_2$ + $3\,H_2O$	
propane	C_3H_8 + $\mathbf{5}\,O_2$ → $3\,CO_2$ + $\mathbf{4}\,H_2O$	
butane	C_4H_{10} + $\mathbf{6.5}\,O_2$ → $\mathbf{4}\,CO_2$ + $\mathbf{5}\,H_2O$	
pentane	C_5H_{12} + $\mathbf{8}\,O_2$ → $\mathbf{5}\,CO_2$ + $\mathbf{6}\,H_2O$	
hexane	C_6H_{14} + $9.5\,O_2$ → $6\,CO_2$ + $7\,H_2O$	

   c. List three patterns you observe for the 6 balanced equations.

2. Examine the carbon compounds in the table below. All carbon compounds combust, except for CO_2.

Combustion Reactions of Carbon and Carbon-Containing Molecules		
methane	CH_4 + $2\,O_2$ → CO_2 + $2\,H_2O$	
methanol	CH_4O + $\mathbf{1.5}\,O_2$ → CO_2 + $2\,H_2O$	
carbon	C + $1\,O_2$ → CO_2	
carbon monoxide	CO + $0.5\,O_2$ → CO_2	
carbon dioxide	CO_2 + $0\,O_2$ → no reaction	

   a. Balance the equation for methanol in the table. Draw the structural formula for CH_4O, methanol.

   b. What patterns do you notice in the amount of oxygen that reacts?

   c. Do you think you can form CO_4? Why or why not?

102   Unit 5: Fire

3. The following table shows the equations for the potential reaction of different metals and salts with oxygen.

Combustion Reactions of Metals		
magnesium	Mg + $0.5\,O_2$ → MgO	
magnesium oxide	MgO + $0\,O_2$ → no reaction	
magnesium chloride	$MgCl_2$ + $0\,O_2$ → no reaction	
titanium	Ti + $1\,O_2$ → TiO_2	
titanium oxide	TiO_2 + $0\,O_2$ → no reaction	
titanium fluoride	TiF_4 + $0\,O_2$ → no reaction	

a. What patterns do you notice?

b. What is the charge on the magnesium atom in MgO?

c. What is the charge on titanium in TiO_2?

d. Why doesn't MgO combust?

e. Do you think that salts combust? Explain your thinking.

4. When carbon-containing molecules combust, a flame is visible. When metals combust, there is a glow, but no flame. Try to explain this observation. (Hint: Molecules enter the gas phase easily. Metals and metal oxides are solids.)

Making Sense:
What are the products of the combustion of carbon-containing molecules?
What are the products of the combustion of metals?
What are the products of the combustion of salts?

If you finish early…
Write a balanced equation for the reaction of $C_6H_{12}O_6$, glucose, with oxygen.

Making Sense Discussion   (10–15 min)

Major goals: Students continue to build their understanding of combustion. They should come away from this discussion with an understanding of two major types of combustion reactions: combustion of carbon-containing molecules and combustion of metals. They should be able to identify the products and be able to identify patterns in the stoichiometry of the reactions. Relate a flame to the presences of glowing hot gases.

6. Discuss combustion equations.

Ask students to help you balance the combustion reactions as you write them on the board.

Discussion goals:

TEACHER GUIDE

Assist students in writing balanced combustion equations. Discuss patterns in the stoichiometry of these equations.

Sample questions:

When ethanol, C_2H_6O, combusts, what is it reacting with? (oxygen)

What are the products of the combustion of ethanol?

How would you write the combustion of ethanol as a chemical equation?

$(C_2H_6O + O_2 \rightarrow CO_2 + H_2O)$

Balance the equation.

When sodium combusts, what is it reacting with?

What are the products of the combustion of sodium?

How do you figure out the formula for sodium oxide?

How would you write the combustion of sodium as a chemical equation?

$(Na + O_2 \rightarrow Na_2O)$ Balance the equation.

Points to cover:

A combustion reaction is a reaction between a fuel and oxygen. In this lesson, we considered carbon-containing molecules and metals as fuels. When carbon-containing molecules combust, carbon dioxide and water are the products. When metals react with oxygen, metal oxides form. Some examples are given below.

Ethanol: $C_2H_6O + 3O_2 \rightarrow 2CO_2 + 3H_2O$

$2Na + 0.5O_2 \rightarrow Na_2O$ (or $4Na + O_2 \rightarrow 2Na_2O$)

7. Discuss why some reactions produce flames.

Discussion goals:

Help students to consider what causes a flame. Discuss the connection between flames and species in the gas phase.

Sample questions:

What are the products of combustion of carbon-containing molecules?

Are the products gases, liquids, or solids?

Do the reactants (carbon-containing molecules) enter the gas phase readily?
  Explain your thinking.

What do you think causes a flame?

What are the products of the combustion of metals?

Are the products of metal combustion gases, liquids, or solids?

Do metals enter the gas phase readily?

Why do you think is there no flame when a metal combusts?

Points to cover:

Each combustion reaction is actually a complex series of smaller steps. The fuel breaks down to smaller compounds and may move from the solid to the gas phase as it combusts. These gases are the cause of flames. Flames are simply highly heated gases giving off light.

One type of combustion reaction considered in this lesson is the combustion of carbon-containing molecules. The combustion products are gaseous carbon dioxide and water vapor. Since the carbon-containing molecules evaporate readily, the combustion reaction takes place in the gas phase. The flame consists of hot gas molecules moving rapidly. The flame has a visible color because some of these gas molecules are emitting light.

A second type of combustion reaction is the combustion of metals to form metal oxides as products. Both the metal reactants and the metal oxide products are solids. Since the metals do not enter the gas phase, the combustion reaction is confined to the surface of the solid. Consequently, a glow is observed, but no flame.

Check-in   (5 min)

8. Introduce the Check-in exercise.

Write the following exercise on the board for students to complete individually.

- Pick a substance that combusts from the list below and write the balanced chemical reaction.

    Ar   Al   C CH_4O

9. Discuss the Check-in exercise.

Get a sense of the level of understanding by taking a vote, collecting students' work, or asking students to defend their choices.

Discussion goals:
Make sure students know how to write balanced chemical equations for combustion reactions.

10. Wrap-up

Assist the students in summarizing what was learned in the class.
- The products of the combustion of carbon-containing molecules are carbon dioxide and water.
- Flames are gases emitting light. Flames are the result of gases that are produced *during* a combustion reaction.
- In a combustion reaction, the longer the carbon chain in the fuel, the more oxygen it reacts with.
- The products of the combustion of metals are solid metal oxides. These reactions usually do not produce enough gases to support a flame.

Homework
11. Assign homework.

Homework – Investigation II – Lesson 3

1. Predict which substance will combust, 10 g of calcium, Ca, or 10 g of calcium chloride CaCl$_2$? Explain your reasoning. Write the balanced combustion reaction for the substance that combusts.

2. Propane (C$_3$H$_8$) is a fuel used in camp stoves.

   a. Draw the structural formula for propane.
   b. Write the balanced chemical reaction for the combustion of propane.

3. Hydrogen is a fuel used in the space shuttle.

   a. Write the balanced chemical reaction for the combustion of hydrogen.
   b. Write the Lewis dot structures for the reactants and product of this reaction.

4. When iron (Fe) rusts it reacts with oxygen.

   a. Balance the following reaction for the formation of rust.

   $$\underline{\quad} \text{ Fe } + \underline{\quad} \text{ O}_2 \quad \rightarrow \quad \underline{\quad} \text{ Fe}_2\text{O}_3$$

   b. What is the charge on the Fe ion? Explain your reasoning.
   c. Do you think rust is a combustion reaction? Explain your reasoning.

5. The chemical composition of sugar is C$_{12}$H$_{22}$O$_{11}$.

   a. What are the products when sugar is burned as fuel?
   b. Write the chemical reaction for the combustion of sugar.

TEACHER GUIDE

All-A-Glow

Name: _____

Period: ____ Date: _____

Purpose: The goal of this lesson is to allow you to examine chemical equations that describe combustion reactions.

Answer the following questions:

1. Molecules containing carbon and hydrogen are called alkanes. Alkanes react with oxygen to produce carbon dioxide, water, and a fire.

    a. Draw the structural formulas for C_2H_6, ethane, and C_4H_{10}, butane.

    b. The reactions for methane, ethane, and hexane are balanced in the table. Balance the reactions for propane, butane, and pentane. (Solve for one unit of the fuel in each case.)

Combustion Reactions of Alkanes	
methane	$CH_4 + 2\,O_2 \rightarrow CO_2 + 2\,H_2O$
ethane	$C_2H_6 + 3.5\,O_2 \rightarrow 2\,CO_2 + 3\,H_2O$
propane	$C_3H_8 + O_2 \rightarrow CO_2 + H_2O$
butane	$C_4H_{10} + O_2 \rightarrow CO_2 + H_2O$
pentane	$C_5H_{12} + O_2 \rightarrow CO_2 + H_2O$
hexane	$C_6H_{14} + 9.5\,O_2 \rightarrow 6\,CO_2 + 7\,H_2O$

    c. List three patterns you observe for the six balanced equations.

2. Examine the carbon compounds shown in the table bellow. All carbon compounds combust, except for CO_2.

Combustion Reactions of Carbon and Carbon-Containing Molecules	
methane	$CH_4 + 2\,O_2 \rightarrow CO_2 + 2\,H_2O$
methanol	$CH_4O + O_2 \rightarrow CO_2 + H_2O$
carbon	$C + 1\,O_2 \rightarrow CO_2$
carbon monoxide	$CO + 0.5\,O_2 \rightarrow CO_2$
carbon dioxide	$CO_2 + 0\,O_2 \rightarrow$ no reaction

Unit 5: Fire

107

    a. Balance the equation for methanol in the table. Draw the structural formula for CH_4O, methanol.

    b. What patterns do you notice in the amount of oxygen that reacts?

    c. Do you think you can form CO_4? Why or why not?

3. This table shows the chemical equations for the potential reaction of different metals and salts with oxygen.

Combustion Reactions with Metals		
magnesium	Mg + $0.5\,O_2$ →	MgO
magnesium oxide	MgO + $0\,O_2$ →	no reaction
magnesium chloride	$MgCl_2$ + $0\,O_2$ →	no reaction
titanium	Ti + $1\,O_2$ →	TiO_2
titanium oxide	TiO_2 + $0\,O_2$ →	no reaction
titanium fluoride	TiF_4 + $0\,O_2$ →	no reaction

    a. What patterns do you notice?

    b. What is the charge on the magnesium atoms in MgO?

    c. What is the charge on the titanium atoms in TiO_2?

    d. Why doesn't MgO combust?

    e. Do you think that salts combust? Explain your thinking.

4. When carbon-containing molecules combust, a flame is visible. When metals combust, there is a glow, but no flame. Try to explain this observation. (Hint: Molecules enter the gas phase easily. Metals and metal oxides are solids.)

Making Sense
What are the products of the combustion of carbon-containing molecules?
What are the products of the combustion of metals?
What are the products of the combustion of salts?

If you finish early…
Write a balanced equation for the reaction of $C_6H_{12}O_6$, glucose, with oxygen.

TEACHER GUIDE

Lesson Guide: All-A-Glow
Investigation II – Lesson 3

In order to understand combustion in more depth, it is important to examine the stoichiometric equations for combustion reactions. This lesson examines both the reactants and the products of combustion, and the ratios in which they combine.

ChemCatalyst

Answer the following question:

The following table shows the balanced chemical equations for four combustion reactions.

Substance	Combustion reaction
methane	$CH_4 + 2\,O_2 \rightarrow CO_2 + 2\,H_2O$
ethanol	$C_2H_6O + 3\,O_2 \rightarrow 2\,CO_2 + 3\,H_2O$
glucose	$C_6H_{12}O_6 + 6\,O_2 \rightarrow 6\,CO_2 + 6\,H_2O$
magnesium	$2\,Mg + O_2 \rightarrow 2\,MgO$

- List three patterns you notice.

Activity

Work individually on the questions below, unless otherwise instructed by your teacher.

Answer the following questions:

1. Molecules containing carbon and hydrogen are called alkanes. Alkanes react with oxygen to produce carbon dioxide, water, and a fire.

   a. Draw the structural formulas for ethane, C_2H_6, and butane, C_4H_{10}.

   b. The reactions for methane, ethane, and hexane are balanced in the table. Create a table similar to the one that follows, in your notebook. Balance the reactions for propane, butane, and pentane. (Solve for one unit of fuel in each case.)

STUDENT GUIDE

Copy this table into your notebook

Combustion Reactions of Alkanes								
methane	CH_4	+	2 O_2	→		CO_2	+	2 H_2O
ethane	C_2H_6	+	3.5 O_2	→	2 CO_2	+	3 H_2O
propane	C_3H_8	+	O_2	→		CO_2	+	H_2O
butane	C_4H_{10}	+	O_2	→		CO_2	+	H_2O
pentane	C_5H_{12}	+	O_2	→		CO_2	+	H_2O
hexane	C_6H_{14}	+	9.5 O_2	→	6 CO_2	+	7 H_2O

c. List three patterns you observe for the six balanced equations.

2. Examine the carbon compounds in the table below. All carbon compounds combust, except for CO_2. (If you want you can copy this table into your notebook.)

Combustion Reactions of Carbon and Carbon-Containing Molecules								
methane	CH_4	+	2 O_2	→		CO_2	+	2 H_2O
methanol	CH_4O	+	O_2	→		CO_2	+	H_2O
carbon	C	+	1 O_2	→		CO_2		
carbon monoxide	CO	+	0.5 O_2	→		CO_2		
carbon dioxide	CO_2	+	0 O_2	→		no reaction		

a. Write out the equation involving methanol, and balance it. Draw the structural formula for CH_4O, methanol.

b. What patterns do you notice in the amount of oxygen that reacts?

c. Do you think you can form CO_4? Why or why not?

STUDENT GUIDE

3. The following table shows the chemical equations for the potential reaction of different metals and salts with oxygen.

Combustion Reactions with Metals				
magnesium	Mg +	0.5	O_2 →	MgO
magnesium oxide	MgO +	0	O_2 →	no reaction
magnesium chloride	$MgCl_2$ +	0	O_2 →	no reaction
titanium	Ti +	1	O_2 →	TiO_2
titanium oxide	TiO_2 +	0	O_2 →	no reaction
titanium fluoride	TiF_4 +	0	O_2 →	no reaction

   a. What patterns do you notice?

   b. What is the charge on the magnesium atom, Mg, in MgO?

   c. What is the charge on Ti in TiO_2?

   d. Why doesn't MgO combust?

   e. Do you think that salts combust? Explain your thinking.

4. When carbon-containing molecules combust, a flame is visible. When metals combust, there is a glow, but no flame. Try to explain this observation. (Hint: Molecules enter the gas phase easily. Metals and metal oxides are solids.)

Making Sense Question:

What are the products of the combustion of carbon-containing molecules?

What are the products of the combustion of metals?

What are the products of the combustion of salts?

If you finish early…

Write a balanced equation for the reaction of $C_6H_{12}O_6$, glucose, with oxygen.

Check-in

Answer the following question:

*   Pick a substance that combusts from the list below and write the balanced chemical reaction.

    Ar   Al   C CH_4O

Homework

Complete the following for homework:

1. Predict which substance will combust, 10 g of calcium, Ca, or 10 g of calcium chloride CaCl$_2$? Explain your reasoning. Write the balanced combustion reaction for the substance that combusts.

2. Propane (C$_3$H$_8$) is a fuel used in camp stoves.

   a. Draw the structural formula for propane.
   b. Write the balanced chemical reaction for the combustion of propane.

3. Hydrogen is a fuel used in the space shuttle.

   a. Write the balanced chemical reaction for the combustion of hydrogen.
   b. Write the Lewis dot structures for the reactants and product of this reaction.

4. When iron (Fe) rusts it reacts with oxygen.

   a. Balance the following reaction for the formation of rust.

   _____ Fe + _____ O$_2$ → _____ Fe$_2$O$_3$

   b. What is the charge on the Fe ion? Explain your reasoning.
   c. Do you think rust is a combustion reaction? Explain your reasoning.

5. The chemical composition of sugar is C$_{12}$H$_{22}$O$_{11}$.

   a. What are the products when sugar is burned as fuel?
   b. Write the chemical reaction for the combustion of sugar.

STUDENT GUIDE

Materials for *Fire* Unit

KIT CONTENTS

2 alcohol burners

matches (1 large box, 16 small boxes, and 1 box long matches)[R]

16 long lab candles[R]

1 short lab candle[R]

3.5 feet of plastic tubing (~1/4-inch diameter)

100 surgical gloves[R]

magnets

pads of steel wool (very fine)[R]

liquid soap[R]

48 plastic spoons

100 large paper clips

300 balloons[R]

20–30 square feet of aluminum foil[R]

1 small lump of clay

300 cotton balls[R]

2 large bags Cheetos®[R]

For up-to-date information about kit contents, please visit www.keypress.com/chemistry .

[R]Consumable; will need to be replenished. This kit includes enough for approximately 10 classes of 32 students, or 2–3 years.

OTHER MATERIALS NEEDED

Common lab ware, lab supplies, chemicals, and other easily obtainable materials have not been included in this kit. They are listed below.

Your classroom should have a sink. Many times you can use hot plates if you do not have Bunsen burners in your classroom.

Safety note: In this unit, we will be burning numerous substances to find out more about fire. The experiments that will be carried out have been selected because they allow for the safe observation of fire. However, caution is always essential when dealing with fire. Review safety procedures with your students (make sure that hair is tied back; do not wear loose clothing; put out small fires by smothering the flames with a fire blanket, water, or sand; etc.)

Have a fire extinguisher available at all times. Remember that a carbon dioxide fire extinguisher and water cannot be used for fires caused by metals. Make sure students know where to get help quickly should the need arise.

A clay pot filled with sand is useful for extinguishing metal fires. Burning items can be plunged into the sand to extinguish them, or sand can be poured onto a fire.

Remind students of the importance of lab safety at the outset of this unit. You may wish to establish that any deviation from procedures will result in loss of lab credit or lab privileges.

Lab Ware

(These quantities are for a class of 32 students)

32 safety goggles (1 per student)

1 pair tongs

16 ring stands with clamps

1 Bunsen burner

8 hot plates

8 scale balances

1 fireproof surface

1 50-mL beaker

24 100-mL beakers

3 500-mL beakers

1 1000-mL beaker

1 250-mL Erlenmeyer flask

2 test tubes (optional)

8 watch glasses

1 glass stirring rod (~1 foot)

16 thermometers

1 funnel

Chemicals

(These quantities are for ten classes of 32 students.)

60 g aluminum powder, Al

40 g copper powder, Cu

250 g iron powder, Fe

10 g magnesium pieces, Mg

20 g magnesium powder, Mg

240 g ammonium chloride, NH_4Cl

650 g barium nitrate, $Ba(NO_3)_2$

240 g calcium chloride, $CaCl_2$

1000 mL 1 M hydrochloric acid, HCl

100 mL concentrated sulfuric acid (18M)

250 g potassium chlorate, $KClO_3$

60 g potassium nitrate, KNO_3

650 g strontium nitrate, $Sr(NO_3)_2$

500 mL rubbing alcohol

800 g sodium chloride, NaCl
(21 teaspoons)

Other Materials

(These quantities are for 10 classes of 32 students)

Baking flour

Water

200 g sugar

10 intact eggshells with the contents blown out

500 mL lamp oil

1 tsp salt

500 mL vinegar (optional, for cleaning steel wool)

160 g cornstarch

1000 paper clips

16 empty tuna fish cans (optional; other containers may be substituted)

16 soda cans with tabs

32 pieces of wire (8–10″) cut from hangers

1 piece of iron wire (6″)

1 9-volt battery

2 wires with alligator clips

1 package non-filter cigarettes (optional; paper or tissue paper may be substituted)

1 dollar bill (optional)

1 dowel or broomstick

1 clay flower pot with sand

1 piece of chalk

1 wide, shallow plastic dish

1 tray (for minimal clean-up)

1 250-mL plastic container (plastic cup or bottle is fine)

String (~1 foot)

Tape

Masking tape

Scissors

Elmer's glue or glue sticks

1 sheet poster paper or butcher paper for Fire Concept Map

LESSON-BY-LESSON MATERIALS GUIDE

An attempt has been made to provide the specialty items needed for each lab. However, common lab ware, chemicals, and other easily obtainable materials have not been included in this kit.

Materials listed here are for a class of 32 students. For consumable items, one kit contains enough for 10 classes of 32 students.

Safety note: In this unit, we will be burning numerous substances to find out more about fire. The experiments that will be carried out have been selected because they allow for the safe observation of fire. However, caution is always essential when dealing with fire. Review safety procedures with your students. Make sure that hair is tied back. Do not wear loose clothing. Put out small fires by smothering the flames with a fire blanket, water, or sand.

Have a fire extinguisher available at all times. Remember that a carbon dioxide fire extinguisher and water cannot be used for fires caused by metals. Make sure students know where to get help quickly should the need arise.

Lesson I-1: Fired Up

Preparation: Prepare some soapy water in the shallow plastic dish. Use enough soap so that blowing into it with a straw produces soap bubbles.

If you do not have Bunsen burners, you may use a lighter.

Tape a long candle to the end of a dowel or broomstick. Inflate a balloon ahead of time or ask a student in class to inflate it for the final demonstration.

Materials Provided in Kit	Other Materials Needed
1 long kitchen match	Student worksheet
1 small box matches	32 safety goggles (1 per student)
1 large lab candle	1 Bunsen burner
1 taper candle attached to dowel	1 methane gas outlet with tube attached
1 alcohol burner	1 ring stand
1 cotton ball	1 watch glass
1 pair gloves	1 100-mL beaker with 50 mL rubbing alcohol
1 balloon	2 100-mL beakers
Liquid soap	1 pair tongs
	1 fireproof surface
	1 glass stirring rod (~1 foot)
	4.5 g potassium chlorate, $KClO_3$
	10 mL concentrated sulfuric acid (18 M)
	20 g sugar
	1 piece of iron wire (~6 in. long)
	1 plastic cup or bottle (~250 mL)

(continued)

Lesson I-1: Fired Up (continued)

Materials Provided in Kit	Other Materials Needed
	1 tray (to minimize clean-up)
	1 wide, shallow plastic dish
	water
	tape

Lesson I-2: Hot and Cold

Preparation: Set up $CaCl_2$, NaCl, and NH_4Cl in beakers or other suitable containers at each table.

Materials Provided in Kit	Other Materials Needed
24 plastic spoons	Student worksheet
	32 safety goggles (1 pair per student)
	8 100-mL beakers
	24 beakers for salts
	24 g ammonium chloride, NH_4Cl
	24 g calcium chloride, $CaCl_2$
	24 g sodium chloride, NaCl
	8 thermometers
	water

Lesson I-4: Now We're Cookin'

Preparation: Students will skewer a Cheeto puff on the end of a straightened paperclip and burn it in this calorimetry experiment. They will need a holder for the burning Cheeto. Taping the other end of the paperclip to an empty tuna can works well; other stands can be improvised depending on what you have available.

You may wish to devise one calorimetry setup at the front of the room as an example for students to follow. You may also wish to have some open windows or other arrangement for increased ventilation.

Materials Provided in Kit	Other Materials Needed
1 bag Cheeto® puffs (at least 16 pieces)	32 safety goggles (1 pair per student)
2–3 boxes of matches	Student worksheet
	16 ring stands with clamps
	16 thermometers
	16 empty tuna fish cans (or other containers)
	16 soda cans with tabs

(continued)

Lesson I-4: Now We're Cookin' *(continued)*

Materials Provided in Kit	Other Materials Needed
	1 roll aluminum foil
	100 paper clips
	2 L water
	tape

Lesson II-1: No Smoking Zone

ALERT: To complete this activity, a cigarette will be lit and "smoked" by an apparatus in class. Thus, you should make sure that students are not asthmatic or allergic to cigarette smoke prior to the demonstration. Open several windows, if possible.

Preparation: Construct your "smoking machine" before class. Use a plastic soda bottle with a cap. Carefully cut a small hole in the bottle cap and insert the tubing through the hole. Use clay around the base of the tubing to completely seal the hole in the cap. Insert an unfiltered cigarette into the other end of the tubing. Place a few cotton balls inside the bottle. Keep at least one fresh cotton ball for comparison. Following the directions in the Teacher Guide, practice using the machine before using it in class.

Inflate the CO_2 balloon (filling it with your breath is fine).

Materials Provided in Kit	Other Materials Needed
1 piece of 6-in. plastic tubing (Cut one 6-in. piece from the plastic tubing; save 3 feet for Lesson II-6)	32 safety goggles (1 pair per student)
Small lump of clay to seal hole in cap	1 500-mL beaker with 50 mL water, labeled "water"
5–6 cotton balls	1 500-mL beaker with 50 mL isopropanol (rubbing alcohol), labeled "isopropanol"
Matches	1 500-mL beaker with 25 mL water mixed with 25 mL isopropanol, labeled "50/50 water/isopropanol"
1 balloon (blow up and label "CO_2")	1 pair tongs
1 long candle	1 package unfiltered cigarettes, or cigarettes with the filters cut off, 1–2 cigarettes per class (optional; paper or tissue paper rolled to the size of a cigarette may be substituted.)
	1 clay flower pot with sand (for putting out paper fire)
	3 paper strips, ~3 × 11 in. each
	dollar bill (optional, for effect; strip of paper may be used instead)
	1 dowel or broomstick to attach taper candle to
	1 plastic 2-L bottle with cap
	tape

Lesson II-2: You're Fired!

Preparation: Fill a clean empty alcohol burner with lamp oil.

Materials Provided in Kit	Other Materials Needed
1 alcohol burner	32 safety goggles (1 per student)
1 small piece of steel wool	1 watch glass
1 candle (Bunsen burner may be used instead)	1 thermometer
matches	50 mL lamp oil
12-in. piece of aluminum foil	50 mL vinegar (optional, to clean steel wool)
	1 teaspoon of salt
	1 piece of chalk

Lesson II-5: Sparklers

Preparation: Cut the wire hangers using wire cutters (provided in the Alchemy kit). Set up the various powders and compounds at each table or at stations around the room. DANGER: DO NOT USE POTASSIUM PERCHLORATE IN THIS EXPERIMENT.

Materials Provided in Kit	Other Materials Needed
8 surgical gloves	32 safety goggles (1 per student)
	2 large beakers
	8 hot plates
	8 balances
	8 watch glasses
	6 g aluminum powder, Al
	65 g barium nitrate, $Ba(NO_3)_2$
	4 g copper powder, Cu
	25 g iron powder, Fe
	2 g magnesium powder, Mg
	16 g potassium chlorate, $KClO_3$
	65 g strontium nitrate, $Sr(NO_3)_2$
	16 g cornstarch
	40 mL water
	32 pieces of wire (8–10 in.) cut from hangers

Preparation for next day: If the sparklers are not too thick or too runny, they will air-dry overnight if stuck upright in a beaker. If they need some assistance, a fan or a blow dryer on a low setting, or a warm (not hot) oven, can be used to dry them. Hollow out an eggshell for each class (see Preparation for II-6).

Lesson II-6: Kablooie!

Preparation: Retrieve the sparklers that students made in Lesson II-5. Inflate a balloon.

You will also need a hollow eggshell for each class, to demonstrate the spectacular combustion of methane. Use a pin to make a small hole (less than 1 cm in diameter) in the top and bottom of an egg, break the yolk, and blow out the contents into a bowl. Rinse out the shell and allow it to dry.

Cut out and assemble a fire tetrahedron (see handout).

You may also wish to "load" the surgical tubing with flour before class.

Materials Provided in Kit	Other Materials Needed
Surgical tubing or long pipette	Sparklers from Lesson II-5
Candle (Bunsen burner may be used instead)	32 safety goggles (1 per student)
Matches	1 funnel
Balloon (filled with air)	1 ring stand
	Natural gas used for Bunsen burners (to fill eggshell)
	Intact eggshell with contents blown out
	Baking flour
	String
	Push pin
	Fire Map on butcher paper from Lesson I-6
	Scissors
	Elmer's glue or glue sticks
	Candle attached to wooden dowel or broomstick

Lesson III-1: No Going Back

Preparation: Fill a balloon with hydrogen by reacting magnesium in hydrochloric acid. Place about 100 mL of 1 M HCl in a 250-mL Erlenmeyer flask. Put magnesium pieces (~1.0 g) inside a balloon. Put the balloon over the mouth of the flask without allowing the magnesium to fall into the HCl. Once the balloon is securely sealed around the flask, shake the deflated balloon to allow the magnesium to drop into the flask. The reaction between the magnesium and HCl generates H_2 gas, which inflates the balloon to about 1 liter. Once the reaction is complete, tie off the balloon. Since the balloon is less dense than air, be sure to tie a string onto the balloon so that you do not lose it. Hydrogen balloons can be made the morning before their use. They don't "save" well overnight.

Set up water electrolysis. Put about 800 mL of water in a 1000-mL beaker. Add about 15 teaspoons salt. Stir until all the salt dissolves. Fold two 8 cm × 8 cm pieces of aluminum foil so that you have two flat strips about 8 cm × 1 cm. Hang the two strips from the side of the beaker. Do not allow the two strips to touch one another.

(continued)

Lesson III-1: No Going Back (continued)

Materials Provided in Kit	Other Materials Needed
1 balloon (to fill with H_2 gas)	1 250-mL Erlenmeyer flask
Matches	1 1000-mL beaker
2 pieces of aluminum foil (~8 cm × 8 cm)	2 test tubes (to collect gases)
	1 ring stand with ring
	100 mL 1 M hydrochloric acid
	1 g magnesium pieces
	800 mL salt water w/60 g salt (15 teaspoons)
	1 9-volt battery
	2 wires with alligator clips
	String (~1 foot)
	Candle attached to wooden dowel or broomstick

Preparation for next day: Prepare a sign for each class, with "Fire" written in invisible ink (see Preparation for III-2).

Lesson III-2: Fire Starter

Preparation: For each class, prepare a sign that says "Fire" in invisible ink for the demonstration. In a small beaker, add potassium nitrate (about 6 g) to 20 mL of water until you have created a saturated solution. Use a glass stirring rod or a cotton swab to write the word "Fire" on the paper. Make sure that you use cursive or continuous writing so that the letters are all connected to each other. Use plenty of solution and make the lines thick. Put a small mark on the paper where the word begins, for later reference. Let the paper dry at room temperature.

Materials Provided in Kit	Other Materials Needed
2 magnets	32 safety goggles (1 per student)
Matches	1 50-mL beaker
	1 stirring rod
	~6 g potassium nitrate, KNO_3
	20 mL water
	1 piece of uncoated paper (~8.5 × 11 in.)
	Masking tape

Lesson III-4: Ashes to Ashes

Materials Provided in Kit	Other Materials Needed
Student worksheet	Fire Concept Map

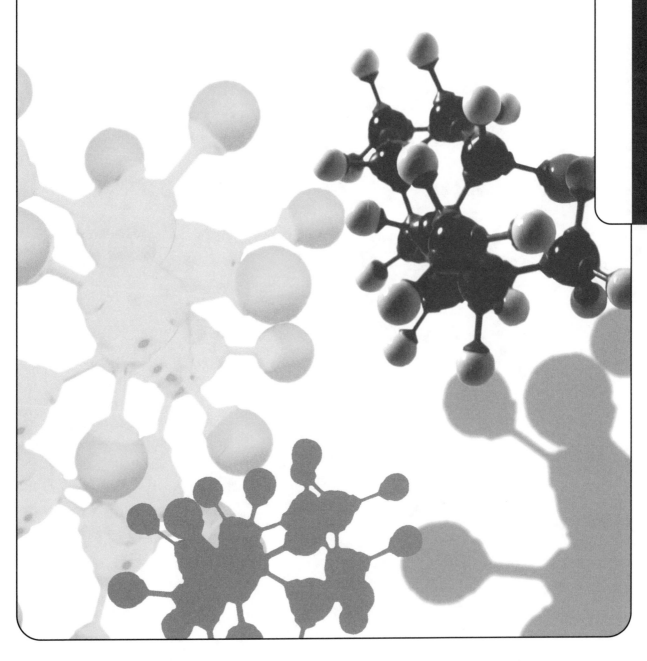

ADDITIONAL
TEACHING RESOURCES

Additional Teaching Resources

WEB-BASED TEACHING RESOURCE CENTER

Teaching resources for *Living By Chemistry*, Preliminary Edition are available online at www.keypress.com/chemistry .

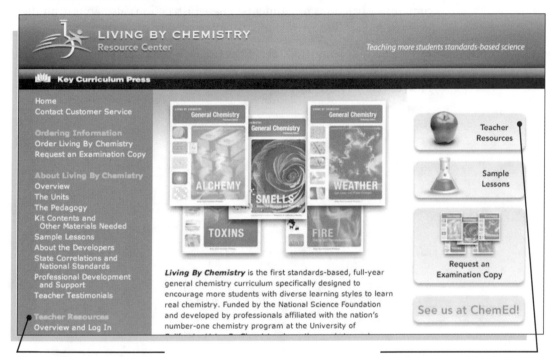

To access these resources, you will need a standard login.
Username: lbc
Password: catalyst

www.keypress.com/chemistry

Resources include

- Quizzes, tests, and exams with answer keys

- Lab assessment options

- Answers to homework exercises

- Ready-to-use classroom presentations in PowerPoint™ format

The Web site also offers up-to-date information about the program. Please visit the site for updates.

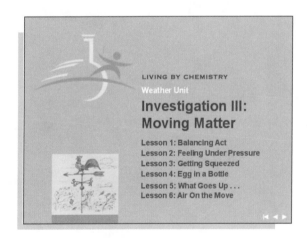

Create a Table Card Sort

The cards printed on the following pages are for use with the Create a Table lesson, beginning on page 6 of this sampler. To try this activity, make single-sided copies of the cards, preferably onto card stock, and cut them out. The last card contains instructions for students. These instructions also appear on the student worksheet. Each group of students will need a set of cards.

Eight full-color laminated sets of the Create a Table cards are included in the *Alchemy* lab kit and they may also be purchased separately from Key Curriculum Press at (800) 995-6284.

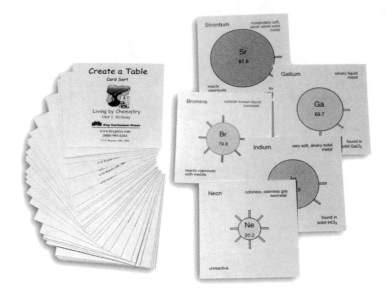

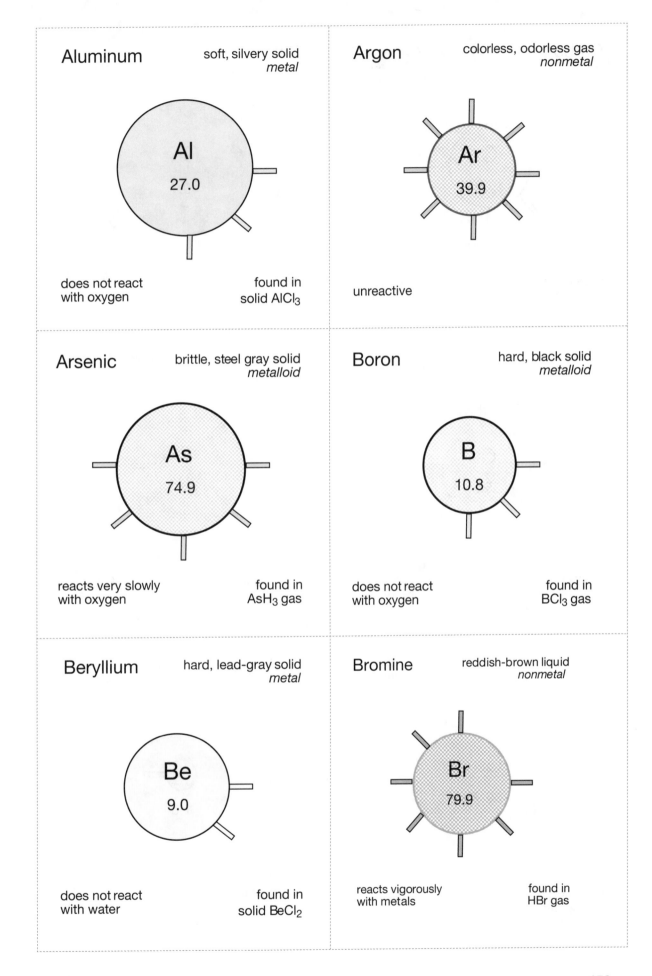

Aluminum   soft, silvery solid
metal

Al
27.0

does not react
with oxygen

found in
solid AlCl$_3$

Argon   colorless, odorless gas
nonmetal

Ar
39.9

unreactive

Arsenic   brittle, steel gray solid
metalloid

As
74.9

reacts very slowly
with oxygen

found in
AsH$_3$ gas

Boron   hard, black solid
metalloid

B
10.8

does not react
with oxygen

found in
BCl$_3$ gas

Beryllium   hard, lead-gray solid
metal

Be
9.0

does not react
with water

found in
solid BeCl$_2$

Bromine   reddish-brown liquid
nonmetal

Br
79.9

reacts vigorously
with metals

found in
HBr gas

Carbon

hard, clear solid (diamond)
nonmetal
or soft, black solid
(graphite)
metalloid

C
12.0

does not react
with oxygen

found in
CH_4 gas

Calcium

moderately hard,
silvery solid
metal

Ca
40.1

reacts
with water

found in
solid $CaCl_2$

Chlorine

greenish-yellow gas
nonmetal

Cl
35.5

reacts violently
with metals

found in
HCl gas

Fluorine

pale yellow gas
nonmetal

F
19.0

explodes upon
contact with metals

found in
HF gas

Gallium

silvery liquid
metal

Ga
69.7

does not react
with oxygen

found in
solid $GaCl_3$

Hydrogen

colorless, odorless gas
nonmetal

H
1.0

explodes in air
when sparked

found in
HCl gas

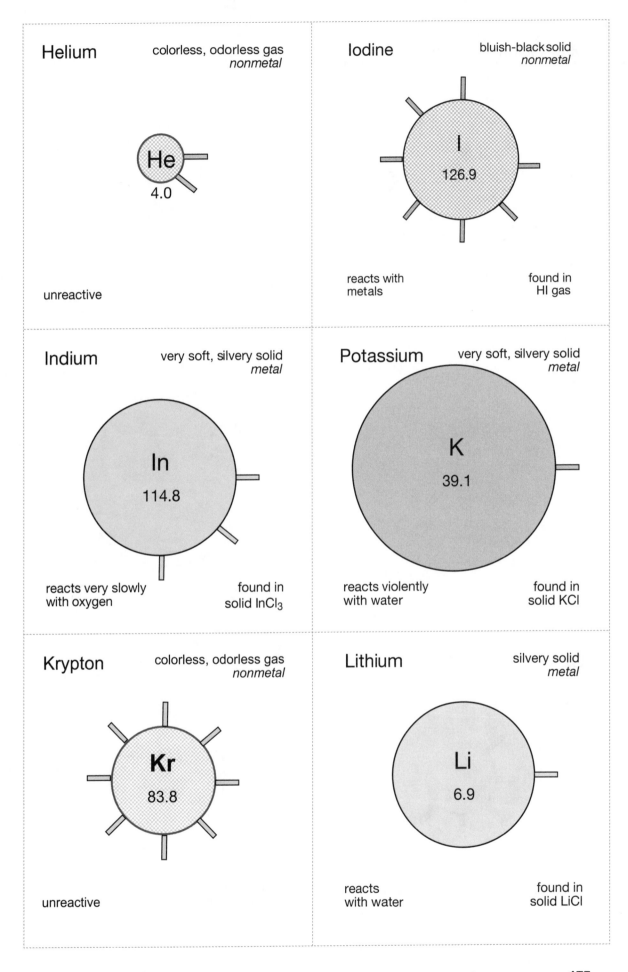

Helium   colorless, odorless gas
nonmetal

He
4.0

unreactive

Iodine   bluish-black solid
nonmetal

I
126.9

reacts with
metals

found in
HI gas

Indium   very soft, silvery solid
metal

In
114.8

reacts very slowly
with oxygen

found in
solid InCl$_3$

Potassium   very soft, silvery solid
metal

K
39.1

reacts violently
with water

found in
solid KCl

Krypton   colorless, odorless gas
nonmetal

Kr
83.8

unreactive

Lithium   silvery solid
metal

Li
6.9

reacts
with water

found in
solid LiCl

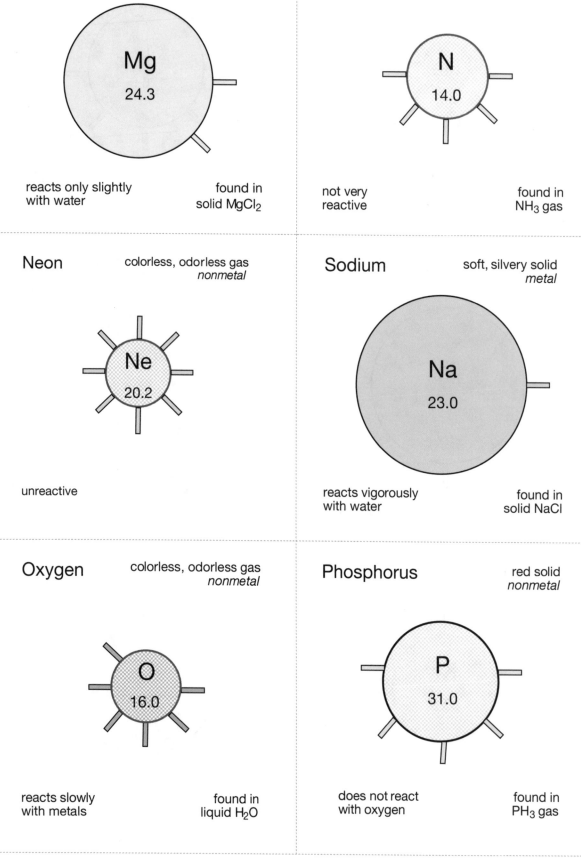

Magnesium

moderately hard,
silvery solid
metal

Mg
24.3

reacts only slightly
with water

found in
solid MgCl$_2$

Nitrogen

colorless, odorless gas
nonmetal

N
14.0

not very
reactive

found in
NH$_3$ gas

Neon

colorless, odorless gas
nonmetal

Ne
20.2

unreactive

Sodium

soft, silvery solid
metal

Na
23.0

reacts vigorously
with water

found in
solid NaCl

Oxygen

colorless, odorless gas
nonmetal

O
16.0

reacts slowly
with metals

found in
liquid H$_2$O

Phosphorus

red solid
nonmetal

P
31.0

does not react
with oxygen

found in
PH$_3$ gas

LIVING BY CHEMISTRY PROGRAM SAMPLER
©2006 Key Curriculum Press

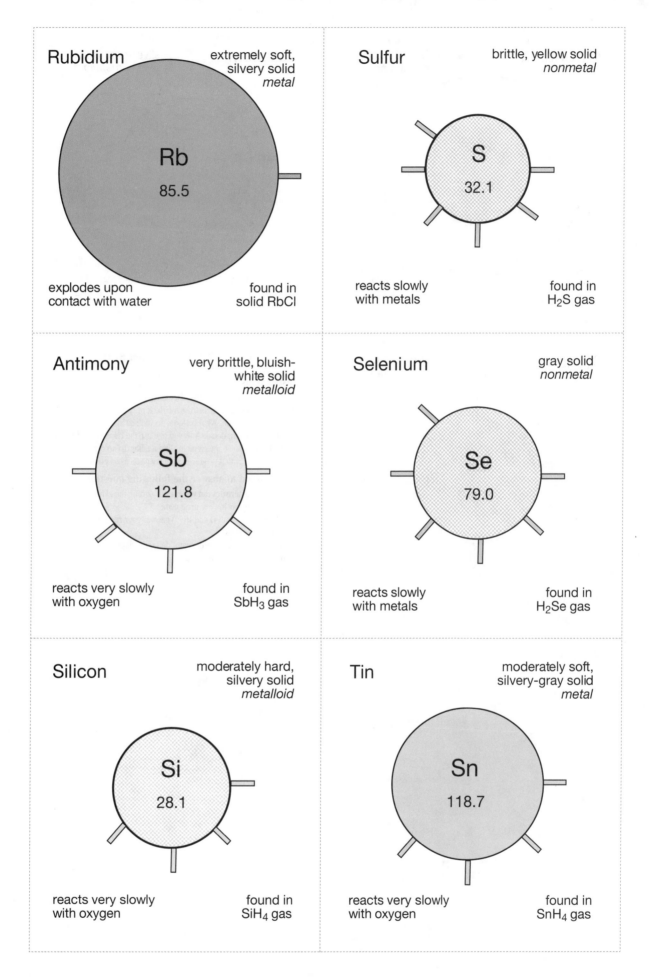

Rubidium — extremely soft, silvery solid *metal*
Rb 85.5
explodes upon contact with water — found in solid RbCl

Sulfur — brittle, yellow solid *nonmetal*
S 32.1
reacts slowly with metals — found in H_2S gas

Antimony — very brittle, bluish-white solid *metalloid*
Sb 121.8
reacts very slowly with oxygen — found in SbH_3 gas

Selenium — gray solid *nonmetal*
Se 79.0
reacts slowly with metals — found in H_2Se gas

Silicon — moderately hard, silvery solid *metalloid*
Si 28.1
reacts very slowly with oxygen — found in SiH_4 gas

Tin — moderately soft, silvery-gray solid *metal*
Sn 118.7
reacts very slowly with oxygen — found in SnH_4 gas

Strontium

moderately soft,
silver-white solid
metal

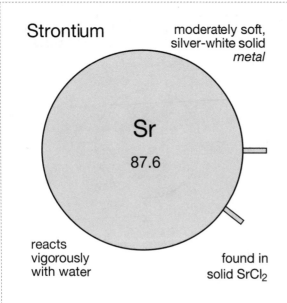

Sr
87.6

reacts
vigorously
with water

found in
solid $SrCl_2$

Tellurium

silver-gray solid
metalloid

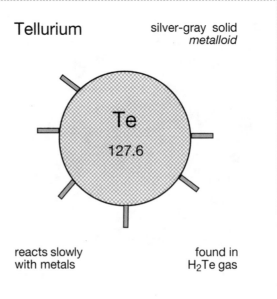

Te
127.6

reacts slowly
with metals

found in
H_2Te gas

Xenon

colorless, odorless gas
nonmetal

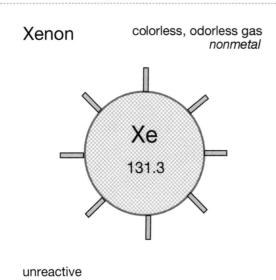

Xe
131.3

unreactive

CREATE A TABLE Card Sort

Instructions

1. Work with your group and discuss each step.
2. Find Be, Mg, Ca, and Sr from the deck of cards, and arrange them in a column. What patterns do you see in this column?
3. With your team, decide how you can use the remainder of the cards to organize the elements into a table. Try to organize them in a way that produces as many patterns as possible.

Be prepared to answer the following questions

1. What characteristics did you use for sorting the cards? What patterns appear in your arrangement?
2. Where did you put H and He? What was your reasoning for their placements?
3. Did you notice any cards that didn't quite fit or that seemed out of order? Explain.